Bicarbonate of Soda

D0493498

100s of Household Uses

Publisher's Note:
All reasonable care has been exercised by the author and publisher to ensure that the tips and remedies included in this guide are simple and safe. However, it is important to note that all uses of bicarbonate of soda should be practised with caution and a doctor's, or relevant professional's, advice should be sought if in any doubt or before any topical or medicinal usage. Neither the editors, the author nor the publisher take responsibility for any consequences of any application which results from reading the information contained herein.

Please note also that the measurements provided in this book are presented as metric/imperial/US-cups practical equivalents, and that all eggs are medium (UK)/large (US) respectively unless otherwise stated.

This is a **FLAME TREE** book
First published in 2011

Publisher and Creative Director: Nick Wells
Project Editor: Catherine Taylor
Picture Research: Katie Pimlott, Catherine Taylor, Harrison Fertig
Art Director: Mike Spender
Layout Design: Jane Ashley
Operations Manager: Chris Herbert

Thanks to Polly Prior, Daniela Nava, Helen Snaith

11 13 15 14 12

3 5 7 9 10 8 6 4 2

This edition published 2011 for Index Books Ltd by
FLAME TREE PUBLISHING
Crabtree Hall, Crabtree Lane
Fulham, London SW6 6TY
United Kingdom

www.flametreepublishing.com

Flame Tree is part of The Foundry Creative Media Co. Ltd
© 2011 this edition The Foundry Creative Media Co. Ltd

ISBN 978-1-84786-980-7

A CIP Record for this book is available from the British Library upon request.

Printed in China

Bicarbonate of Soda

of Soda

100s of Household Uses

Diane Sutherland, Jon Sutherland, Liz Keevill and Kevin Eyres

**FLAME TREE
PUBLISHING**

Contents

Introduction

If you could find a naturally occurring product that you could use as a deodorant, a toothpaste, an exfoliant and an antiseptic, you would be impressed. If you found out that you could use the same substance in the kitchen to make a big improvement to many of your favourite dishes, and then use it to clean out your pots and pans, leaving them grease-free and shining, you would be amazed. And if you were told that exactly the same product could be used to shampoo your pets, clean out your swimming pool, kill cockroaches and relieve insect bites and stings, you'd probably just laugh. Well, there is just such a product, it's been around for thousands of years in one form or another and you can buy it today for pennies.

It All Started in Ancient Egypt

When the ancient Egyptians needed something that would keep mummified bodies dry and free from bacteria on their journey through the afterlife, they found the answer: a sodium compound, occurring naturally in dry lake beds. They called the substance Natron because a particularly good source was an area called Wadi-el-Natrun.

The Egyptians quickly discovered that Natron could also be used as an extremely effective cleaner around the home. Not only that, but blended with olive oil it made soap; in its natural state, it could be used as toothpaste and diluted it made a breath-freshening antiseptic mouthwash.

What was more, they found that Natron burned with a smokeless flame when mixed with castor oil – handy for working in tombs without leaving soot stains. And it was useful in chemical processes such as glass making. All in all a pretty remarkable and versatile substance.

Then Came Pearl Ash and Soda Ash

Over the centuries, the various elements that make up Natron were refined down to more specific substances for individual uses. Late in the eighteenth century, European chemists discovered that another form of sodium, pearl ash, was particularly useful in dramatically speeding up the baking process. However, the production of pearl ash involved the burning of huge amounts of wood to produce ash, and pretty soon wood for pearl ash was in short supply.

In 1791, French chemist Nicolas LeBlanc produced a process for turning common salt (sodium chloride) into soda ash (sodium carbonate), a substance with the same characteristics as pearl ash but without the need to burn vast tracts of woodland.

And Finally There Was Bicarbonate of Soda

American bakers loved soda ash and America was soon producing its own. However, American soda ash could not match the quality of the European product, so American bakers were forced to import large quantities of European soda ash.

All that changed in the 1830s when an American, Austin Church, began experimenting with new ways of producing high-quality sodium bicarbonate. Church, a doctor, developed a process that converted purified sodium carbonate into food-grade sodium bicarbonate.

Giving up his medical practice, Church joined forces with his brother-in-law, John Dwight, moved to New York, and founded John Dwight & Co. Dwight called his sodium bicarbonate 'saleratus' (aerated salt) and was soon selling it to bakers and housewives across the States.

Thirty years later, Austin Church retired from the business a rich man, but he was not finished with sodium bicarbonate.

With his two sons, he set up Arm & Hammer to produce bicarbonate of soda products, and if you go into any chemist or supermarket, you can still find Arm & Hammer toothpaste on the shelves.

You Can Call It:

- bicarbonate of soda (more commonly used in the United Kingdom)
- bicarb
- baking soda (especially if you're American)
- bread soda (again, more common in the United States)
- sodium bicarbonate
- saleratus
- or even, sodium hydrogen carbonate

But Do Not Confuse It With:

- caustic soda (sodium hydroxide)
- baking powder (although it does contain bicarbonate of soda)
- sodium chloride (common salt)
- sodium carbonate (also known as washing soda, or soda ash)

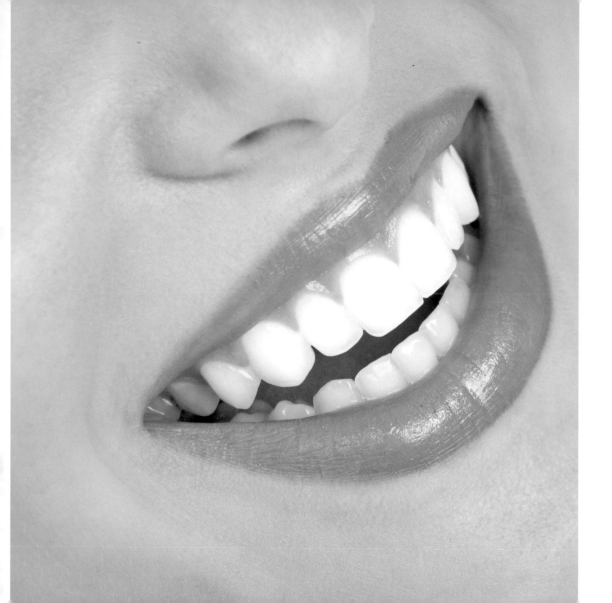

Cleaning

Kitchen

If you glance at the ingredients of most of the cleaning products that live under your sink, you will find a scary list of some truly horrendous-sounding chemicals, many refined from petrol and most of them highly toxic. It does not have to be like that. Bicarbonate of soda used in various ways, neat or diluted, can replace many of those cleaners at a fraction of the price and at considerably less risk to your health. Here's how...

Washing-up

End that Sinking Feeling

For effective, eco-friendly washing-up, simply add 1 tablespoon bicarbonate of soda to a sinkful of hot water and swish in the juice of half a lemon. If you use commercial washing-up liquid, add a couple of tablespoons (or more) of bicarb to the water to boost its action. This is really effective when dealing with greasy dishes.

A Handy Stand-by

Keep a little bowl of dry soda close to the sink. If you encounter a stubborn stain, dip your washing-up cloth, brush or sponge in and scrub – the offending mark will be gone in a jiffy. This not only works on crockery and pans but is also great for glass and plastic cookware.

 CAUTION: Do not use on nonstick finishes.

And Here's Another One

Pop some bicarb into a flour or sugar shaker and keep it by the sink. Use it to sprinkle on stains, then scrub and rinse. Or recycle a plastic scouring powder or talcum powder container and use to hold dry bicarb. Whatever sort of container you use, be sure to label it carefully so it does not get mistaken for food.

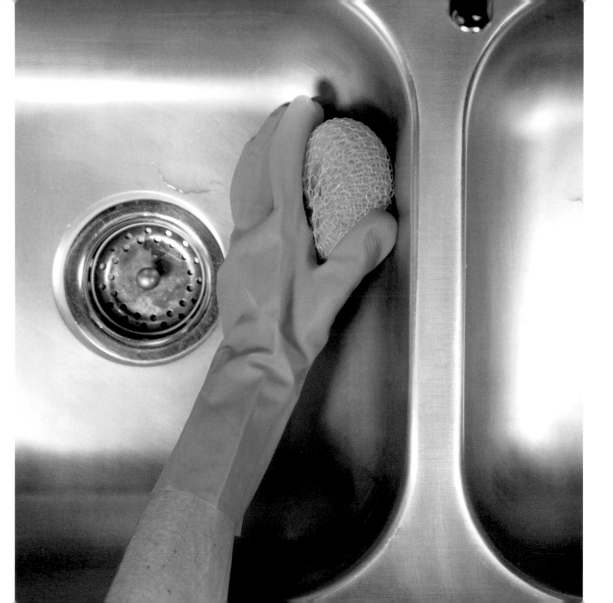

Banish Dirty Dishcloths and Washed-up Sponges

Rinse dishcloths and washing-up sponges after use in a solution of hot water and bicarb to keep them smelling like roses. Occasionally soak overnight for an extra powerful freshen-up. Always dry cloths thoroughly before putting away to keep them hygienic and to avoid the build-up of unpleasant smells.

Up to Your Elbows in Hot Water?

Keep those Marigolds smelling like daisies by sprinkling a little bicarbonate of soda into your rubber gloves each time you take them off. Not only will it absorb any dampness and keep them fresh for next time, they will also glide on and off like a dream.

In the Dishwasher

Why spend money on multicoloured tablets with a ball stuck in the middle for no apparent reason? Make up your own dishwasher powder by putting 2 tablespoons bicarbonate of soda and 2 tablespoons borax into the powder compartment for a full load.

What Is that Smell?

To freshen up a whiffy dishwasher and give it a thorough clean, sprinkle about 150 g/5 oz/¾ cup bicarb into the bottom of the machine and run it empty through a complete hot cycle. Whiffiness over.

Reduce odour build-up between cycles by sprinkling about a handful of soda over the dirty dishes and on to the bottom of your dishwasher.

Clean Plates but Dirty Dishwasher?

For stubborn stains in your dishwasher, use dry soda on a damp cloth or sponge, or mix up a paste with soda and a little water and use like cream cleaner. This is particularly effective for getting rid of those nasty bits that build up around the hinges, rubber door gaskets and in other hard-to-reach areas of dishwashers. Rinse well to remove any residue.

Welcome Home

No one wants to return to a smelly dishwasher after they've been on holiday, so sprinkle a little bicarb into the empty machine and leave the door slightly open. It will be a delight to come back to.

Give Chopping Board Smells the Chop

Chopping (cutting) boards can acquire a less-than-pleasant smell over time, so give yours a regular spring clean by shaking over

about three tablespoons of bicarb and sprinkling with just enough water to moisten. Leave for about 15 minutes to absorb those onion and garlic odours, then rub well with a wet sponge. Rinse in clean water and allow to dry.

Pots and Pans

For those boring things that will not go in the dishwasher, mix together equal quantities of bicarbonate of soda, borax and salt, and use as a scouring powder that will cut through the heaviest grease and grime on pans, grills and baking trays.

Remember to rinse well afterwards, and do not use on nonstick surfaces, such as Teflon, as it will damage them. Soda will also tend to make aluminium pans go darker.

Don't Want to Wash Them Right Now?

Alternatively, soak dirty pots and pans in a basin of hot water with 2 or 3 tablespoons bicarb for about an hour, then scrub them clean with an abrasive scrubber. Again, not for nonstick under any circumstances.

Left a Pan on the Heat and Forgotten About It?

To remove seriously burned-on food, soak the pan in bicarb and water for 10 minutes before washing, or scrub the pot with dry soda and a moist scouring pad. Do not use abrasive cleaners on nonstick housewares.

For a really scorched pan with burned-on food in the bottom, scrape off as much of the debris as you can, then pour a thick layer of baking soda directly into the pan. Moisten the baking soda with a little water, then leave to work overnight. This should loosen the burnt food and you'll just need to scrub clean and rinse. If this still does not work, boil a strong solution of soda in the pan for about 10 minutes. This should soften up the burnt food sufficiently for you to scrub it away with some dry soda on a damp scourer.

What About Nonstick?

For burnt pans that you don't want to scrub, such as nonstick, bring a mixture of water and 250 ml/8 fl oz/1 cup vinegar to the boil in the pan. Remove from the heat and add a couple of tablespoons of soda. Leave to soak overnight then wipe clean and rinse well.

Mugs and Cups

Use a paste of bicarb and water to remove tea and coffee stains from ceramic and melamine cups. Leave the solution in the cup for a while, then rub and rinse. So much more ecological than bleach, and it does not leave an aftertaste.

Surfaces

Work on those Worktops

Wipe your kitchen worktops over with a little dry bicarbonate of soda sprinkled on to a damp cloth or sponge. Rinse with clean water and wipe dry. It is a safe and natural way to keep surfaces clean and leaves them free of taint.

Alternatively, mix up a solution of bicarb and hot water and use this to wash down your worktops, leaving them sparkling and fresh. Rinse and dry as above. Rinse your cleaning cloths or sponges in the same solution and dry thoroughly before putting away.

Freshen up Formica

To erase stains on laminate worktops such as Formica, use a thick paste of bicarb and rub gently until the offending mark disappears. Rinse well. If that does not fix it, squeeze some lemon juice on to the stain, leave for around half an hour, then add a little dry soda to the lemon. Rub with a sponge, rinse clean and dry.

Shining Sinks

Sinks come up sparkling and bright when you clean them with bicarbonate of soda. Use dry on a damp cloth, or make up a paste with soda and water, and use like cream cleaner.

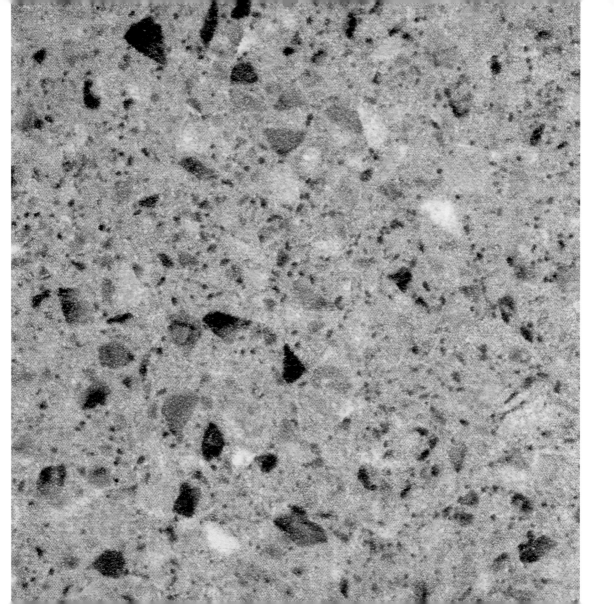

Degrease Lightning

Prevent the build-up of grease in your kitchen sink drain by regularly putting a large handful of dry bicarb down. Do this about once a month for best results, and make sure you do it just before you set off on holiday so there are no nasty drain smells on your return.

Alternatively ...

Keep your sink drain fresh by regularly putting a handful of bicarb down the plug with the same amount of salt, and follow it down with a cupful of boiling water. Leave for about half an hour, then pour down a whole kettleful of boiling water.

Back to White

If your porcelain or white enamel sink has taken on an unpleasant yellowish tinge, get it looking like new by mixing 1 litre warm water with 50 g/2 oz/¼ cup soda and 120 ml/4 fl oz/ ½ cup chlorine bleach. Pour into the sink and leave to soak for quarter of an hour.

 CAUTION: Keep children and pets away while you are doing this. Rinse well until the bleachy smell goes completely.

Dispense with Dirty Waste Disposals

Keep waste disposal units clean by switching the unit on and running hot water through it. While it is still running, add a handful of soda, then keep the tap running until all traces of powder have disappeared.

Appliances

Top Cooker-top Tip

Make a paste of soda and water and use like cream cleaner on a sponge or cloth. Wipe over well to remove all white streaks and rinse with clean water. Alternatively, dampen the oven top or hob thoroughly and sprinkle with baking soda. Leave for half an hour to absorb any grease, then rub clean with a sponge or cloth and rinse well. Wipe over glass hobs with a solution of hot water and bicarb on a sponge or cloth. Tackle any stubborn marks or burned-on food with dry bicarbonate on a wet cloth.

Detox Your Oven Cleaner and Your Oven

Proprietary oven cleaners can be highly toxic, and often contain caustic chemicals that give off dangerous fumes. Not nice. However dirty your oven is inside, bicarb can beat the burned-on grime. Scrape off as much gunge as you can, then mix plenty of bicarbonate with a little water to make a thick paste. Spread this over the sides and bottom of the cold oven and leave overnight. Wipe down the next day with hot water and repeat if necessary. Use dry bicarb on a scourer or sponge to deal with really stubborn areas, rinsing well afterwards.

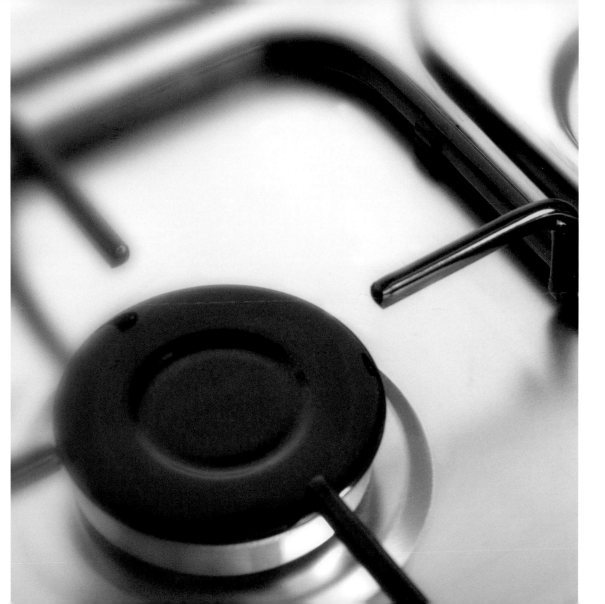

Oven Spritzer

Another way to keep ovens fresh and clean is to spray the interior of the oven with water (a plant or laundry sprayer works well for this), then sprinkle with dry bicarbonate of soda. Spray again to dampen the powder. Give further spritzes with water every hour or so, leave the mixture overnight and then remove with a cloth. Rinse with hot water.

Can't See through that See-through Door?

Glass oven doors can get so splattered with grease and grime that you cannot see through them any more. Get them crystal clear again by scrubbing with bicarb on a damp cloth or sponge. Rinse well and dry.

Microwave Maxi Cleaning

Insides of microwaves can get sprayed with over enthusiastically heated food, leaving them smelling musty, especially when the door is kept closed. To refresh your microwave without leaving any taint or chemical smells, remove the turntable plate if there is one, and wash this in the sink in a solution of hot water and bicarb, then dry.

Wipe over the inside of the microwave with a solution of 1 1/2 pts/1 qt hot water and 4 tablespoons bicarb. Be sure to wring your cloth or sponge out well before wiping, as you don't want water to get into the workings through any little perforations in the inside of the oven.

Tackle stubborn splashes with dry bicarbonate on a damp cloth or sponge and rinse clean. Dry the inside of the microwave and replace the turntable. You can wipe down the outside using the same solution.

More Microwave Magic

Another way to freshen your microwave is to add a couple of tablespoons of bicarb to a half-full cup or mug of hot water (make sure the cup or mug is microwave-safe). Put in the oven on full power and boil for around five minutes. The steam will condense on the inside of the warm oven, loosening splatters of food and leaving it damp. All you need to do is wipe over with kitchen paper, a cloth or sponge, and then dry.

If you must leave your microwave door closed when it's not in use, you can eliminate musty smells by leaving a small open container of bicarb inside. A ramekin, cup or glass yogurt pot is ideal. Just remember to remove it before using the oven and replace it afterwards. Stir the contents each time you do and replace the whole lot every two or three months.

Freshen up Your Fridge

Fridges need a complete spring clean every now and again, especially if something has gone off in there and the smell just will not go away. Remove all the food, shelves, drawers and containers, and wipe them over, along with all surfaces and nooks, with a solution of hot water and bicarbonate on a clean cloth or sponge (an old toothbrush is useful for difficult

crannies, and you can use dry powder on the wet brush for stubborn marks). Rinse and dry carefully before switching the appliance back on.

If you need to leave a fridge switched off, leave the door propped open and pop an open container of bicarb inside.

And Don't Forget the Freezer

If you have to defrost your freezer, wipe it over inside with a solution of 4 tablespoons bicarb to 1 1/2 pts/1 qt warm water. Wipe over all surfaces and dry before turning the freezer back on.

If your freezer is frost-free, it can be tempting to leave it for aeons without cleaning it. But even frost-free freezers appreciate a wipe down inside from time to time. Use the same solution as above.

Clean on the Inside, Clean on the Outside

Kitchen appliances, cookers, fridges, hobs and so on can get amazingly grimy when you're not looking. Wipe the whole lot over regularly using a solution of bicarb and warm water. Use a cloth or sponge and wring out to avoid drips.

Odour Removal

Fridge (Odour) Magnet

Strong cheese is great, but things like that can leave a not-so-great smell in the fridge. Fight back by leaving an open container of bicarb in there to absorb strong smells and eliminate stale odours. Stir the contents regularly and replace every two or three months or you will find the bicarh itself is starting to pong!

A Fresher Salad Crisper

While you're dealing with the fridge, sprinkle the bottom of the salad drawer with a little dry bicarb, then cover with a layer of kitchen towel. Wipe out and change about every three months.

'Vacuum' Your Flask

Get rid of musty or unappetizing smells in a vacuum flask simply by adding a teaspoonful of bicarbonate of soda, filling it with hot water, then leaving it to soak for at least 30 minutes. Empty and rinse well with cold water. And leave the top off if you're not going to use it for a while.

And a Fresher Air Freshener

Fill a small bowl almost to the top (choose a pretty one for added effect) with bicarb and add a few drops of your favourite essential oil. Place anywhere in the house that you need to keep things smelling sweet. Add more oil when the effect starts to wane and replace the whole lot about every three months.

When it is not convenient to have an open bowl around (such as under a sink), cut the feet off a clean old pair of tights and fill the foot with soda. Knot the leg and place this inside the other cut-off foot. Knot this again and put it anywhere you need to eliminate stale smells.

Smelly Hands?

If you have just been chopping onions or garlic, peeling potatoes or handling fish, it can be difficult to erase the smell, even with soap. Bicarb will get rid of the whiff – sprinkle some on to wet hands, rub together well and rinse off.

Onion Odour Eaters

Remove the odour of onion and garlic from wooden or other porous surfaces by sprinkling

some bicarb on to a damp cloth and rubbing it into the surface. Rinse with water, or leave for about half an hour to work for stronger smells.

Plastic Food Boxes

Plastic can absorb strong smells and less-than-appetizing odours can build up in plastic food boxes, especially if they are left closed or were not scrupulously clean when put away. Banish those musty odours by washing the container in a hot water and bicarb solution, then sprinkle about 3 tablespoons of soda into the base of the container and top up with hot water. Put the lid on and shake gently over the sink, then leave for at least a couple of hours to work. Rinse out and wash again as usual.

Seasonal Smells

Musty smells can build up in food containers that are only used during certain times of the year. To prevent this, sprinkle the insides of picnic hampers and cool boxes with a little bicarb before putting away. Wipe out with a damp cloth before using.

Ban Smelly Bins

If your kitchen bin has a nasty smell that just will not go away, make a solution of 200 g/7 oz/1 cup bicarb in warm water. Pour it into the bin and top up with more warm water. Leave it to soak for an hour or two, then empty down a drain or the loo. Make sure you rinse and dry carefully, as bacteria and mould (the cause of bad smells) multiply much faster in warm, steamy conditions.

To prevent smells building up in the first place, sprinkle a little bicarb in the bottom of the bin, then line the base with a couple of sheets of newspaper, before inserting a (biodegradable) bin liner.

If a liner leaks, do not ignore it. Deal with it right away by wiping out the bin with a bicarb solution and rinsing with clean water. Change the newspaper and bicarb regularly, every month or so, to keep things really fresh.

And Bread Bin Mould

Accumulated crumbs in the bottom of a bread crock or bin can easily turn mouldy (ugh). Make sure you tip them out regularly and wipe over the inside of the bin with a solution of bicarb and warm water. Remember to dry the bin out thoroughly and allow it to cool before you place fresh bread in it, or the mould will return.

Bathroom

The bathroom is a place to pamper yourself as well as keep clean, so it is really important that you keep it spotless and germ-free. And because they are often damp and steamy, bathrooms easily attract mildew and mould, and no one wants to stare at those when they're soaking in the tub. As your skin comes into contact with the surfaces, the bathroom is the last place you want to use toxic chemicals for cleaning, so it's time to reach for that tub of bicarb again.

General

All-over Clean

For general bathroom cleaning, including the bath, basin, shower tray and wall tiles, mix a paste of bicarbonate of soda and water and use it like cream cleaner on a damp cloth or sponge.

Get rid of more stubborn stains with dry bicarb on a damp cloth or sponge. Rinse well to remove any streaks and wipe dry.

Sweet Smells

Bicarbonate of soda is perfectly safe for all surfaces, and a couple of drops of essential oil added to the cleaning mix will add extra freshness, so no need for harmful air sprays. Give lemon, pine, tea tree or ylang ylang a try.

Grimy Nonslip Strips?

Nonslip strips are a useful safety measure but difficult to keep looking spotless. You will find they get dingy after a while, which may also make them less effective. The answer? Wet the strips and sprinkle with bicarb. Leave to work for about half an hour, then rub clean with a sponge or cloth and rinse well. Job done.

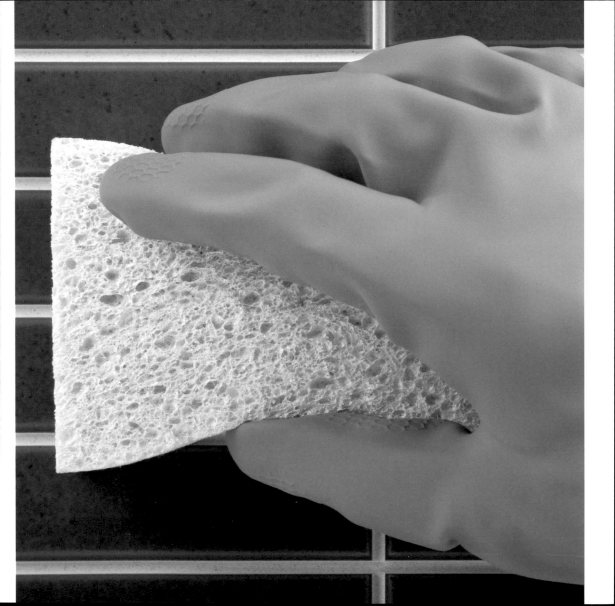

Say Goodbye to Bathroom Odours

Musty smells can lurk in bathroom cabinets and cupboards, particularly under basins where there might be slight leakage from pipes and taps. Keep your bathroom storage areas smelling fresh by putting a small open container of bicarb in each one; this will also help absorb damp.

If you use bicarb as an air freshener in your bathroom, remember to replace it about every three months, but don't throw it in the bin – tip it down the loo, leave for a while, then flush, and you'll get extra value from it.

Leaving an open bowl of bicarb outside the bathroom or by the loo will help to absorb any nasty niffs. Choose a pretty container that goes with the room scheme and, again, replace the contents about every three months. Add a few drops of essential oil for a luxurious touch.

And Banish Musty Bathroom Bins

Bathroom bins can quickly get a bit musty, which is both unhygienic and unpleasant. If your bin is washable, rinse it clean with a solution of bicarb and dry carefully. All bins, whether washable or not, will definitely benefit from a sprinkle of soda in the bottom, covered with a square of kitchen towel. Tip or wipe out the powder about every three months and replace the paper as necessary.

Bathroom Hardware

Tap Dancing

If you live in a hard-water area, there's a good chance that there is a limescale deposit around the base of your taps – annoying and hard to shift. Not any more – mix some bicarbonate of soda with a little vinegar and brush the fizzy paste on the offending areas. Leave for about half an hour, then rub gently and rinse clean. This works equally well for all hard-water marks in the bathroom, including on glass shower screens.

Keep Your Drains Running Clear

Regular drain cleaning helps minimize unpleasant smells and guards against blockages. Keep your drains clear by mixing two parts bicarb, two parts salt and one part white vinegar and pouring the solution down the plugholes in your bath, basin and shower. Put in the plugs and leave the mixture to froth away merrily for about half an hour, then run the hot tap to rinse. Easy.

Bath Time

Bicarbonate of soda is safe for cleaning all types of baths, but you probably don't always want the bother of cleaning it straight after you've had a soak. No problem: just sprinkle a couple of tablespoons of bicarb into the bathwater before you pull out the plug and your bath will be as clean as you are.

Shower Power

Any type of shower cubicle can be cleaned quickly and effectively with bicarb-and-water paste. If you're feeling lavish, add a little shampoo or shower gel to the mix for a nice smell.

If you live in a hard-water area, you will know that glass shower doors soon get covered in annoying water marks. Don't worry: just wipe the glass over with a little white vinegar on a cloth or sponge and then rub over with a sprinkle of bicarb. Rinse well and then squeegee the glass for streak-free spotlessness you can be proud of. Keep a squeegee in the shower and get all the family to have a quick run over the surfaces with it after each shower. It will cut down on cleaning enormously.

Clear Your Head

Another common problem in hard-water areas is blocked showerheads – not what you need first thing in the morning. If you are able to remove the head, soak it in a mixture made up of 250 ml/8 fl oz/1 cup vinegar with 3 tablespoons bicarb in a suitable container slightly bigger than the head, until the limescale has completely dissolved.

If for any reason you cannot remove the head to clean it, tip the bicarb-and-vinegar mixture into a strong plastic bag, a little bigger than the showerhead, and tie or tape it in place over the head so it is in close contact with the frothy solution. Leave in place for the mixture to work its magic for about half an hour, then remove the bag and rinse the head with clean warm water. Run water through the head to clear any residue and you will soon be singing in the shower again.

Curtains for Mouldy Shower Curtains

A mouldy shower curtain is the last thing you want flapping around when you're trying to get clean. Soak yours regularly in warm water with about 5 tablespoons bicarb (the bath is a good place for this). Give it a slosh around a few times, then squeeze out and hang to drip dry. Always let shower curtains dry flat after washing, otherwise you will find the mould will be back again. Not nice.

For mouldy patches or stubborn stains, make up a thick paste of bicarb and rub clean using a sponge, cloth or old nailbrush or toothbrush.

Machine-washable shower curtains can be laundered on the gentle setting with about 100 g/3½ oz/½ cup bicarb or half that amount of mild washing powder and the same of bicarb. Throw in a couple of towels to absorb suds and water, and help wash the curtain (which is probably waterproof!) more effectively. Adding 250 ml/8 fl oz/1 cup white vinegar to the rinse cycle will help.

 CAUTION: Do not use the spin cycle or you could end up with a permanently creased curtain (and shower curtains are not usually designed to be ironed). Drip

dry instead in an airy place (outside is ideal) or hang it back up in the shower and leave the window open.

Be Kind to Your Loo

Proprietary loo cleaners use harsh, caustic chemicals such as bleach, which can stain your clothes if it splashes, harm your hands and, of course, end up in the water supply. Be gentle with your loo: instead, sprinkle some bicarbonate of soda around the bowl, leave for about half an hour and then scrub with a toilet brush and flush.

Remove more stubborn marks by sprinkling the toilet bowl with bicarb and pouring a dash of white wine vinegar on top. Use a toilet brush to scrub the bowl clean with the bubbling froth that results.

Sitting Comfortably?

Any kind of toilet seat will come up like new when wiped over with a mild solution of bicarb. Don't forget to do the underneath of the seat and lid too. When you have finished, tip the rest of the solution down the loo for a really thorough clean.

Flushed with Success

Every few weeks, sprinkle some bicarb in your cistern. Leave overnight, then flush the next day for a sparkly clean loo.

Furnishings & Surfaces

Our homes are under constant attack from dust, germs and bugs, let alone pets, children and the occasional party. Sofas, chairs, curtains and cushions all have to put up with a lot of mistreatment, walls are often on the receiving end of grubby hands (tiny or otherwise!), and floors can suffer from spills and simply from people walking around in shoes designed for the pavement or the park, not the parquet. Bring all these surfaces back to life with a few handfuls of bicarbonate of soda.

Furniture

Back to White

White and light painted furniture can quickly lose its original brightness. Bring it up like new by wiping faded items over with a solution of 2 tablespoons bicarb to 1 1/2 pts/1 qt warm water. Scrub black marks off chair legs with a paste of bicarb and water. Rinse and dry.

Ring the Changes

Get rid of cup rings and other heat marks on wooden furniture by gently rubbing with a bicarb-and-water paste. If this doesn't work, add a pea-sized blob of toothpaste for extra abrasion. Make sure you do not get the furniture wet as this may cause further stains. Wipe over, dry carefully and, finally, buff with some furniture polish.

Dry-clean Your Three-piece Suite in Fifteen Minutes

To give any fabric upholstery a quick dry-clean, sprinkle the surfaces generously with bicarb (use a flour shaker or empty talc container). Leave for about 10 or 15 minutes to absorb any stale odours, then vacuum up all the powder.

Degrease

Annoying grease marks on cloth upholstery can be made to vanish by sprinkling with one part bicarb and one part salt. Simply brush the mixture in lightly with an old toothbrush, then leave overnight to absorb the stain. Vacuum up the next day.

Alternatively ...

Make a paste of 1 tablespoon water and 3 tablespoons bicarb and rub into stains on fabric seat covers – an old toothbrush works well. Leave the mixture to dry, then brush carefully with a clean dry brush and vacuum away any residue.

 CAUTION: It is best to test the mixture first on an inconspicuous area to check that the fabric is colourfast and that it will not affect the surface of the fabric or leave water marks.

A Cure for Under-the-weather Leather and Vinyl

Leather and vinyl furniture often has a textured finish that can trap grease and dirt, leading to a generally grimy effect. Revive the finish by mixing 1 tablespoon bicarb with 250 ml/8 fl oz/ 1 cup warm water and use on a cloth or sponge to wipe down the furniture. Blast more

stubborn grime with a paste of bicarb and water. Rubbing this into all the nooks and crannies will leave everything sparkling clean. Wipe over well with clean water on a sponge or cloth and dry.

Metal Furniture

Grimy metal furniture can be brought back to life by cleaning with a paste of bicarb and water on a cloth or sponge. Rinse clean and dry carefully to prevent rust formation.

Rust-buster

If rust does appear on metal furniture, wipe over the affected area with a paste made from 1 tablespoon bicarb and a few drops of water on a damp cloth, then polish with a piece of baking foil and wipe down with a clean damp cloth and dry carefully with kitchen paper.

Lacklustre Laminates?

Any laminate surfaces that are looking a bit off-colour will respond well to being spruced up with a damp cloth or sponge sprinkled with bicarb. Rinse and dry.

Floors & Walls

All Floors

Looking for an excellent general-purpose hard-floor cleaner that costs pennies? Try dissolving a handful of bicarb in a bucket of warm water and use as normal on a floor mop. Wring out well to avoid saturating the surfaces, then stand back and admire the finish.

Good for Wood

Water can damage parquet and other solid wooden floorings. To remove any offending water marks, take a damp cloth or sponge and sprinkle a little bicarb on to it then rub carefully. Wipe over with a well wrung-out cloth or sponge and dry carefully.

 CAUTION: Do not get the floor wet as this will obviously cause further problems.

Lifeless Lino, Vinyl or Cork Flooring?

Unsightly scuff marks from dark shoe soles on lino, vinyl flooring and cork tiles will vanish when rubbed with a paste of bicarbonate of soda and water.

Crying Over Spilt Ink?

It's no use crying over spilt ink – you just have to do something about it. For spills on hard surfaces, wipe up as much ink as possible, then sprinkle with bicarb. Leave for a few minutes to absorb any residue, then add more powder and rub at the stain with a wet cloth or sponge. For really stubborn stains, moisten the powder with a little vinegar instead of water.

On the Carpet

The sooner you deal with any stain, the better your chance of successfully removing it. This applies just as much to carpets as anything else. If you spill wine or drop grease on your favourite Axminster, sprinkle the area with bicarbonate of soda (or one part bicarb and one part salt), brush in gently and leave it to absorb the stain for several hours. Sweep up the powder with a clean brush, then dab with a solution of bicarb and warm water. Leave to dry completely, then vacuum the area.

Routine Meaures

If you are going to vacuum the carpet, it will be much fresher afterwards if you sprinkle it with bicarb first. Leave it for about 10–15 minutes, or better still overnight, to work, then vacuum the carpet until all traces of powder have disappeared.

Preventative Strike

Before having a new carpet laid, vacuum the bare floor carefully, then sprinkle it all over with bicarb to help keep your carpet smelling fresh. Make sure you tell the fitters not to sweep up the powder, though.

That's Sick

Cleaning up vomit is never a pleasant job, but bring bicarbonate of soda to the rescue and you can ensure there are no lingering nasty memories. Get your rubber gloves on (sprinkling a little bicarb inside them for freshness and ease of removal) and scrape up

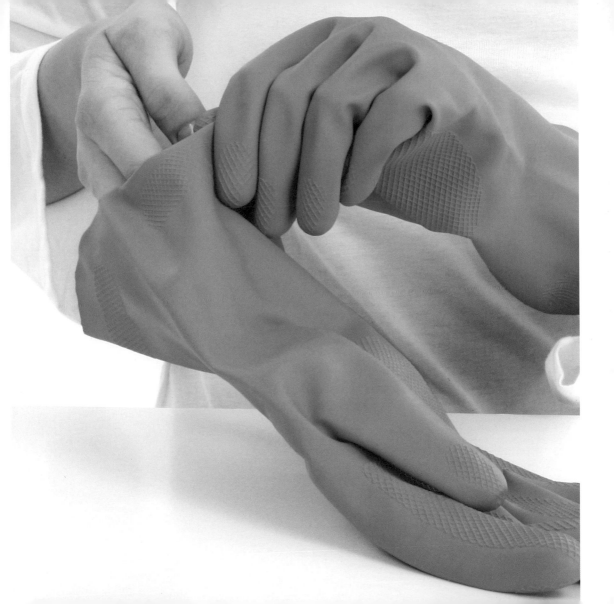

as much mess as possible, then cover the area with a layer of bicarb. Leave for a few minutes, then scrub from the edges towards the middle with a sponge or brush moistened repeatedly in clean warm water. Sprinkle the area again with bicarb and leave to dry, then vacuum up the powder.

The antibacterial action of bicarb will deal with any residual odours, even on carpet. This process also works well with urine.

If you cannot clear up vomit immediately because you're in the middle of an emergency, throw some bicarb over the affected area to curb nasty smells and deal with it as soon as possible.

Removing 'Artwork' from Walls

To clean marks, such as those from crayon, pencil, marker pens or grease, off washable walls, make a solution of bicarb and warm water and wipe over. If any children's masterpieces or other stains prove hard to shift, sprinkle dry bicarb on a damp cloth or sponge and scour gently.

 CAUTION: It is a good idea to test out on an inconspicuous area of wall first, just in case the finish is not as washable as you thought.

Washable Wallpaper

Rub stains on washable wallpaper with a paste of bicarb mixed with a little water. However, bear in mind there is washable and washable, so in case your paper is 'wipeable' rather than 'scrubbable', always test on a less noticeable part of the wall first.

Revive Dull Tiles

Wipe dull, lifeless tiles over with a solution of bicarb, and see them spring back to their old shiny selves. Use the powder dry on a wet cloth or sponge for more stubborn stains and to clear soap residue. When you have finished, rinse with clear water and wipe or squeegee dry.

Defeat Grubby Grouting

Grubby grouting will never show your bathroom in its best light. Revive stained tile grout with a thick paste of bicarbonate of soda and an old toothbrush. Make sure you rinse the whole area clean when you have finished to remove any white streaks.

Alternatively, for general discolouration on tile grout, make up a fizzy paste using two parts bicarbonate of soda and one part vinegar or lemon juice. Apply the paste to the grout with an old toothbrush, leave for 10 minutes and then rinse the whole lot off with warm water.

If your grout has really stubborn stains that simply refuse to budge, it's time to use a big gun. Moisten the bicarb with bleach instead of water (but not more than one part bleach to three parts bicarb) and scrub the grout with this, rinsing well afterwards and wiping.

 CAUTION: If you do decide to use the bleach mixture as a last resort, you will need to take a few simple precautions: open the window and wear rubber gloves and old clothes, wash any splashes off your skin immediately, keep the mixture away from children and pets and, of course, make sure you dispose of any leftover mixture safely.

Windows

First Class Glass

It is amazing how dirty windows can get without our noticing. Get in the habit of washing your windows regularly with a wet cloth or sponge sprinkled with a little bicarbonate of soda. Rinse over with clean water and polish with a crumpled sheet of newspaper until dry, which will help remove any water marks or smears.

And don't forget the frames. Wipe them down with a mixture of one part vinegar to two parts bicarb. This is especially effective if condensation has left areas of mildew.

Beautiful Blinds

Washable blinds (do make sure they are washable) will come back to their just-bought glory once you have doused them in a bath of warm water and 250 ml/8 fl oz/ 1 cup bicarbonate of soda.

Wipe away the dirt with a sponge or washing-up brush (give the cords a quick scrub too), then rinse down using the shower head or take outside and give a gentle blast with the garden hose. Hang the blind up to dry.

 CAUTION: Place an old towel in the bottom of the bath first to prevent the blind scratching or marking it.

Mirror, Mirror on the Wall

Like windows, mirrors can quickly lose their lustre. Bicarb is great for bringing up grubby mirrors so they gleam.

See yourself in a new light by sprinkling a little bicarb on a damp sponge or cloth and rubbing over the mirror. Wipe over with a clean cloth or sponge and fresh water, then polish with a crumpled newspaper for a sparkling, smear-free finish.

General Household

Wherever the water goes when we empty a sink or bath, or flush the toilet, you can send it swiftly on its way, making sure there are no lingering, unpleasant smells, with a dash of bicarbonate of soda. Not especially glamorous maybe, but vitally important for our health and wellbeing. And bicarbonate of soda is great at neutralizing the smell of things around the house, such as shoes, trainers and pets that are making their presence felt a bit too much.

All-purpose Cleaner

Drain Training

If the water isn't draining away as fast as it normally does, or you suspect you have got a blockage, it's time for action. Pour about 100 g/3½ oz/½ cup bicarbonate of soda down the sink, swiftly followed by about 120 ml/4 fl oz/½ cup of white vinegar. Put the plug in while the chemicals froth up in a rather satisfying way. Leave bubbling away for a couple of hours, then pour down a kettle of boiling water. Used regularly, this combination will break down the fatty acids that block drains and help to keep them smelling fresh as a daisy.

Going on Holiday?

Don't forget to put a handful of bicarb down all plugholes in the home before you go away for any longer than a couple of days. It will prevent your neighbours suffering any unpleasant bad drain smells while you're gone, especially in hot weather, and it will make your place much nicer for you to come home to.

Sceptical about Septic Tanks?

If you have a septic tank, don't worry. Bicarb is best for keeping it in tiptop shape. In fact, it positively helps the process of breaking down waste by keeping the contents at the correct pH level, which favours the right sort of bacteria. Bicarb also helps protect the fabric of the tank from corrosion that can be caused by too-acid an environment. So, if you don't have mains drainage, pop a handful of soda down the loo each week and give your tank a healthy treat.

Metal Cleaning

Silver Lining

Stop using unpleasant chemicals on your silverware. Instead, banish tarnish by applying a paste of water and bicarb, or dry bicarb on a damp cloth or sponge. Rinse or wipe with clean water and dry.

Adding a tiny squeeze of mild detergent or shampoo to the bicarb paste when you are cleaning the family silver will give you even more cleaning power. Rub, rinse and polish dry.

If you are lucky enough to have some large, flat pieces of silverware, cut a potato in half and dip into a tub of bicarb. Rub over the surface of the silver, then polish clean with a soft cloth.

The Family Jewels

You can clean several small items of solid silver jewellery together by placing them in a suitably sized flat glass dish lined with a piece of slightly scrunched-up aluminium foil or baking foil.

Arrange the jewellery on the foil, ensuring each piece is in contact with it. Sprinkle the pieces with bicarb, then boil a kettle of water, let it cool slightly and pour over the items until they are all submerged. Turn the items over gently so all surfaces are in contact with the foil in turn. The tarnish will transfer to the foil. Magic.

Get Stainless Steel and Chrome Really Stainless

Bring any stainless steel and chrome surfaces back to life with a sprinkle of bicarb on a damp cloth or sponge. If you have 'brushed' stainless steel surfaces, these have a grain, so be sure to rub in the right direction. Wipe away any residues with a clean damp cloth and buff up to a good-as-new shine with a dry cloth.

Brassed off with Your Copperware?

Bring copper, brass or bronze up like new by mixing three parts bicarbonate of soda and one part lemon juice or white vinegar and cleaning with a soft cloth. But remember that if you are polishing brass, you must make sure it is uncoated or it will not work.

Copper Bottoms?

Clean copper pans by cutting a plump fresh lemon in half and dipping it in bicarb. Rub the surface of the pan with a circular motion. Wipe clean with fresh water and dry carefully.

The Ring of Confidence

Rings can get clogged up and grimy from daily contact with oily deposits from skin, hand cream, soap or other cosmetics. Bring your precious jewels back to pristine condition by soaking them in a solution made from 1 tablespoon bicarb thoroughly mixed with a cup of tepid water.

 CAUTION: Do not use this method on pearls, or jewellery where the gems may be glued in place rather than set in claws. If in doubt about submerging jewellery, rub gently with a bicarb paste using an old, soft toothbrush. Rinse and dry carefully before putting away.

Odour Removal

Whiffy Carpets?

Carpets and rugs can get very musty, especially if you have pets. The solution is to sprinkle with bicarbonate of soda the night before you vacuum. Sleep sound in the knowledge that the bicarbonate is hard at work absorbing smells. Vacuum as normal the next day.

Get It off Your Chest

Some people like that evocative *Lion-the-Witch-and-the-Wardrobe* smell that comes from old wooden furniture, but no one likes the way the smell can transfer to clothes, making them feel less than fresh. Banish the smell to Narnia by sprinkling some bicarbonate of soda in the bottom of the piece of furniture, leave it overnight and vacuum out in the morning, then sprinkle in a bit more and place clean paper over the top.

Banish that dreadful smell of decaying mothballs that hangs about in some ancient furniture by filling a pomander with bicarbonate of soda, or equal quantities of bicarb and borax, and hanging it up in the wardrobe.

Alternatively ...

If you're feeling creative, or short of cash, make a sachet from old clean tights, or fill a small container and stand it in the corner of the cupboard. Well-washed plastic food containers, or small sturdy cardboard boxes with holes punched in the lid, filled with bicarb, are ideal. Having a lid makes it less likely to get tipped over and spilt.

Blanket Measures

If blankets smell musty when you take them out of storage, sprinkle them with bicarb and fold them up. Leave them overnight, then shake out in the garden and hang in the sunshine, or give them a 20-minute twirl in the tumble dryer on a low setting.

A Bed of Roses

Mattresses can get fusty, especially if not used for a while – not a pleasant place to lay your head. Get rid of the fug by removing the mattress cover if there is one, dusting the mattress with bicarb and leaving it overnight before vacuuming it up. Turn the mattress over and repeat the process.

Add a further sprinkling of bicarb before you put back the mattress cover (it goes without saying you've washed it!) and make up the bed. Pillows will benefit from the same treatment.

Butt Out

If you're a smoker, sprinkling a layer of dry bicarbonate of soda in your ashtrays offers a double whammy. Not only does it absorb unpleasant smoky odours, but it also works as a fire-extinguisher, so helps put smouldering butts out. Unfortunately it won't do anything about cleaning up your lungs.

Fusty Footwear

Let's face it, trainers, shoes, boots and slippers can start to smell horrible if not taken care of properly. Avoid embarrassing stink-foot by sprinkling a couple of tablespoons of bicarb into each shoe and shaking carefully to distribute throughout its whole length. Tip down the loo before wearing (the powder, that is, not the shoes) to keep your toilet smelling fresh as well.

For a potentially less messy solution, make sachets to go in your shoes to keep them fresh. The simplest way is to pour some bicarb into the foot section of old tights, knot at the ankle and trim. Pop one in each shoe when you're not wearing them.

Sprinkling a bit of bicarb on your feet (and in your socks) before you put your shoes on will help deal with footwear odour at the source if you are prone to this problem.

Game, Set and ... Smell

Sports and swimming bags can smell truly
ghastly if they are not kept aired and free
of dirty or wet kit. Avoid the shame of
stinky kit-bagitis by sprinkling the inside of
the emptied bag with bicarb. Leave it
overnight, then vacuum out. Sprinkle a bit
more in the bottom and leave it there to
absorb moisture and bad odours.

If you are a member of a sports club with your
own locker, it will stay fresh and much more civilized if
you pop a small open container of bicarb in the corner.

Upstairs, Downstairs

Attics, lofts, basements and cellars often have a characteristically damp, fusty smell. Tights
to the rescue again: cut the legs off and fill the foot section with bicarb. Add another layer of
tights to make double thickness and then knot firmly. Use the rest of the leg to make a loop
and attach it to a rafter, beam or hook. It might look odd (actually, it does look odd) but it
does the trick.

Love Your Luggage

Suitcases, trunks, travel bags and other items of luggage often get stored for months at a
time and can get a musty smell, especially when left in the loft or attic. When you get your

luggage down for a trip, sprinkle a little bicarb inside the day before you pack, leave overnight and vacuum thoroughly in the morning.

If spots of mildew appear on superior-looking leather luggage that has been left in storage or damp conditions, bring it back to its rightful position in life by gently rubbing the affected areas with a paste of bicarb and water. Dry off with a clean cloth, leave to dry thoroughly, then buff up with a suitable coloured polish.

Tame that Tome

Old books can be yet another source of unpleasant musty odours. Use an old librarian's trick once you're sure the book is dry, and sprinkle lightly throughout with bicarb. Leave the book for a few days, then shake out the powder. The result: a book it is a pleasure to get your nose stuck into again.

Accidental water spills can cause the pages of books to buckle and stick together, but don't despair. Dry out wetness (or dampness) on paper by sprinkling the affected pages with bicarbonate of soda and leave the book to dry in a sunny window or the airing cupboard, turning the pages regularly.

Pets

We all love our pets; that's a given. But sometimes they can smell less than fragrant, and occasionally they disgrace themselves when we're not looking. Wet dogs, cat litter and hamster cages all have their unique aroma and it's not one that adds to the pleasure of relaxing at home. Reach for the bicarbonate of soda and they will soon be your best friends again.

Bathing & Floors

Washing, Wet or Dry

Give your pooch a quick dry-clean by sprinkling him (or her, of course) with a little bicarb and brushing it through the coat. Probably best done in the garden, this will absorb grease and remove those doggy smells you do not want to smell.

For a wet wash, add a couple of tablespoons of bicarb to the water when you bath your dog. This solution can be used by itself, when it will also help heal any skin conditions, or with a dash of shampoo. You will not need to use as much shampoo as usual, as the addition of bicarbonate of soda makes it work much more effectively. Rinse thoroughly after shampooing.

A Cleaner Vacuum Cleaner

Every time you change the bag in your vacuum cleaner pop in 1 tablespoon bicarb. And if it's the bagless type just drop the bicarb into the dirt collection chamber. This will quickly curb that less-than-fresh smell that can linger when the cleaner starts to get full, especially if you have a dog in the house.

Whoops

When man's best friend leaves a nasty niff on the carpet, sprinkle on some bicarbonate of soda. Leave it for half an hour, then vacuum up. The bicarb will also help stop those doggy smells gathering in your vacuum cleaner.

Worse Than Whoops

After a night on the tiles, or swallowing too much fur, your cat can sometimes be sick and will seldom, if ever, use the litter tray. Wherever your cat decides to throw up, a solution of bicarbonate of soda and water will not only clean up the mess, it should also remove any stains and unpleasant smells.

The Worst

Should one of your pets leave something unspeakable on the carpet, scrape up or dry up as much as you can, then scrub the whole area with a solution of bicarb and warm water. Sprinkle the area liberally with more bicarb and leave overnight, vacuuming up any residue the next day.

Getting rid of that telltale smell quickly will discourage the little darlings from soiling the same spot repeatedly.

Pets' Accessories

Kitty Litter

With the best will in the world, most us do not manage to clean out litter trays quite as often as we should. Luckily, bicarbonate of soda will absorb those telltale cat litter odours if you plan ahead.

Next time you change your cat's tray, cover the bottom with a layer of bicarb and then sprinkle the cat litter on top. Be generous with the bicarb: the best arrangement is a ratio of around one part bicarb to four parts of litter. Add a little extra bicarb each time you remove anything between changes.

And So to Bed

Pets' beds and bedding can get dirty and start to smell. Freshen up your pet's sleeping arrangements by sprinkling with bicarb, leaving for at least 10 minutes, then vacuuming. Be sure to remove Fido first!

Blanket the Smell

Pet blankets can be home to some particularly powerful odours that even a good hot wash won't always shift. Bring your blankets back to smell-free cleanliness by adding 100 g/3½ oz/ ½ cup bicarb to the detergent, either for machine or hand washing.

White-collar Pets

Dogs' collars and leads can get grimy and greasy over time, so scrub with a solution of bicarbonate of soda and warm water. Nylon collars and leads can be soaked in the solution if you prefer. This also works well on those plastic toys that your pet loves gnawing on.

How Clean Is Your Cage?

Next time you clean out a small pet's cage, sprinkle a layer of bicarbonate of soda on the bottom before covering with a layer of newspaper, then bedding. Tank-style cages can be washed out with a solution of bicarb and warm water and rinsed in the bath or shower.

Feathered Friends

Birds can get salmonella and other similar gastric problems if you do not keep their feed hoppers and drinking containers spotless.

Wash these carefully in a hot water and bicarb solution and rinse and dry carefully before replacing. Use the same solution, rather than strong-smelling detergent, to clean out the cage.

Thanks for All the Fish (Tank)

Fish do not like it if their water becomes too acidic, so test the water with a special kit regularly and add a pinch or so of bicarb to correct the level if necessary.

And when you need to clean the tank, bicarb and water in paste form is perfect for scrubbing it out. It is safe and leaves no taint.

 CAUTION: Make sure you rinse the tank out carefully before refilling.

Laundry

Washing

Sparkling clean, sweet-smelling laundry is one of life's pleasures. But as detergent manufacturers compete to bring us cleaner, brighter, whiter, fresher laundry at lower temperatures, they simply add more and more chemicals to their powders. In the long run, this leads to fabrics not lasting as long and an increasing number of people finding they are allergic to washing powders. Bicarbonate of soda is a simple, inexpensive solution – its mildly alkaline content means it dissolves dirt and grease, leaving laundry soft and clean.

Cleaning Clothes

Painless Stain Removal

Any type of stain is always best dealt with before it has time to dry because once the mark starts to set it will become much more difficult to remove. Because of its unique formulation, bicarb is great for many types of stain removal as well as for general laundry.

 CAUTION: Some fabrics that are optimistically described as machine or hand washable are really not suitable for this kind of washing at all. To be on the safe side, always test stain removal techniques on an inconspicuous area of the material first, especially if it is your absolute favourite item of clothing.

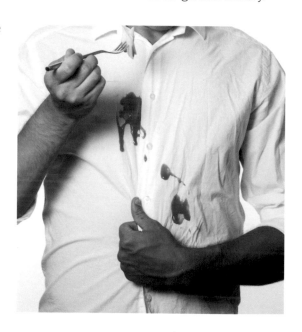

Blood Simple

Blood stains are always difficult to remove, as anyone who watches crime TV shows will tell you. Bicarbonate of soda, however, will normally remove blood from washable clothing.

For small areas of blood, dampen the stain with cold water, but not hot water as this sets blood stains, making them far more difficult to remove. Rub bicarb into the stained area with an old toothbrush or clean nailbrush. Leave the bicarb paste on to soak for a while and repeat the process if necessary. Launder as normal.

Blood Not So Simple

For larger areas of blood, add 200 g/7 oz/1 cup bicarb and 250 ml/8 fl oz/1 cup white vinegar to a bucket of cold water. Swirl the mixture around well and then add the bloodstained clothing. Leave for several hours – overnight is ideal – then launder as usual.

Vomit and Urine

There is no nice way to say this, but vomit and urine are acidic. The first thing to do is thoroughly rinse any stains under a cold tap as soon as possible after the event to avoid lasting damage to the fabric. Sprinkle the area with bicarbonate of soda and leave for about

half an hour before washing. Bicarbonate of soda will neutralize the acid and prevent it from permanently damaging clothing.

For larger areas of vomit, scrape off as much as you can, then soak the item or items in a bucket of cool water to which you have added 200 g/7 oz/1 cup bicarb.

Couldn't Face Cleaning It at the Time?

If acidic stains or spills have dried on to clothes, do not soak them in water as the acid will be reactivated and will start to harm the clothes. This explains why clothes that have had acid splashed on them go into the wash intact and come out with holes.

Instead, mix some bicarb with a little water and spread on to the stain. This will neutralize the acid before it has time to start eating into the fabric. Leave the bicarb on for an hour or two, then wash as normal.

This also works for other acid spills such as lemon, orange and other citrus juices (which, although they are good for you, are acidic), acidic toilet cleaners and battery acid. You can also try this for wine spills and other fruit stains.

No Sweat!

Scrub those unpleasant perspiration stains from washable clothing with a nailbrush and a paste of bicarb and water, leave for at least an hour to work effectively, longer for bad stains, then wash as normal. Not only will this remove the stain, it will get rid of the smell too. Phew ...

Rusty Clothing?

Rust is not just a problem for suits of armour. Rust stains can also appear on suits made of cloth and other clothing. Soak the area with lemon juice, then sprinkle generously with bicarb. Leave to work overnight, then rinse off and wash as normal.

Ease off Grease

Grease stains can be so frustrating. They just seem to come back wash after wash. Don't despair: sprinkle fresh grease splashes with bicarb and leave for an hour or two to work, then drip a little water on to the powder and scrub gently. Rinse off and wash as normal.

For dry-clean only clothes, just sprinkle the dry powder on as above, leave to work and brush off with a clean soft brush.

Ballpoint Blobs?

Ballpoint pen or other types of ink can be removed from leather clothing, but you need to act fast. Sprinkle a little bicarb on the stain, leave for a couple of hours to absorb the ink, then brush off and buff up the leather.

Grey-collar Worker?

Greasy, discoloured shirt collars and grimy cuffs will get short shrift if you scrub them with a paste of bicarb and water and a nailbrush. Leave the mixture to soak for half an hour or so, then wash as normal.

For severe stains, try drizzling a little white vinegar on to the soaking mixture. The resulting froth will deliver the extra grease-busting power that you need.

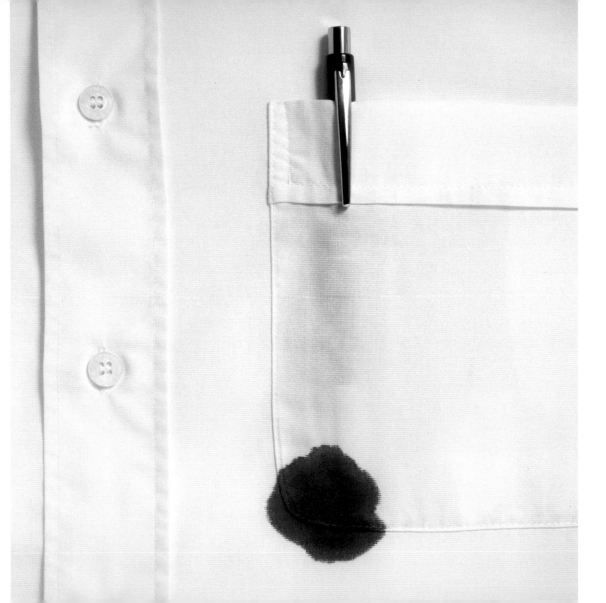

Persuade Suede to Come Clean

Suede is all too easy to stain and expensive to dry-clean. Before you head to the cleaners, try gently rubbing dry bicarb into stains on suede, using a soft brush. Leave it to set, then brush off carefully. It makes sense to try this on an inconspicuous part of the garment before you proceed to ensure it does not have any adverse effects.

New Baby, New Clothes

New baby clothes often contain a dressing that stiffens the fabric so it looks good in the shop. Perversely, this can cause soreness and rashes on delicate baby skin. Rinse this unnecessary chemical concoction out of the clothing before the bambino wears it by washing the offending garment in warm water and a little mild detergent to which you have added 100 g/3½ oz/½ cup bicarb.

Some adults also find their skin is irritated by these fabric dressings, so just follow the same instructions.

Happy Nappies

Disposable nappies can leave a dubious ecological legacy. If you are using the traditional towelling alternative, soak the dirty nappies in bicarb and water overnight. This not only curbs unpleasant smells, but means that you can use less detergent in the machine when it comes to washing them.

In the Swim

Remove damaging chlorine or salt from your bathing costume after a trip to the pool or swimming in the sea by soaking the garment for an hour or two in warm water to which you have added a handful of bicarb. Wash as usual and your swimwear will keep looking good for much longer, and it will not have that worryingly bleachy smell when you next put it on.

Tune up Your Washing Machine

Adding half a cup of bicarbonate of soda to a load of washing will improve the performance of your usual liquid detergent, helping with the removal of stains and grease. And, as bicarbonate of soda acts as a very effective water softener, you will find you get the same results with considerably less detergent than usual.

What's more, your white clothes will come out looking whiter, and brightly coloured clothes will keep their colour longer. All for a few pennies.

Reduce Rashes

If, in common with many people, you are allergic to commercial washing powders and find they irritate your skin, try using 200 g/7 oz/1 cup bicarbonate of soda in your wash instead. This is also ideal for getting baby clothes lovely and soft.

A Handy Hand-washing Tip

You can hand wash delicate articles in warm water with 50 g/2 oz/¼ cup bicarb and about a tablespoon of hand-washing laundry liquid. Wash and rinse as usual, adding a tablespoon of bicarb to the final rinse as a softener.

 CAUTION: Before you wash anything delicate, you should pay particular attention to the washing instructions that came with the garment.

A Soul Mate for Your Sole-plate

The sole-plate of your iron can easily get marked and hard water will soon block up the steam holes. Remedy this by unplugging the iron and leaving it to cool, then rub over the plate with bicarb-and-water paste on a sponge or cloth. Wipe over with a clean damp cloth and buff dry.

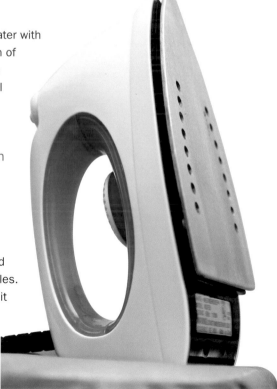

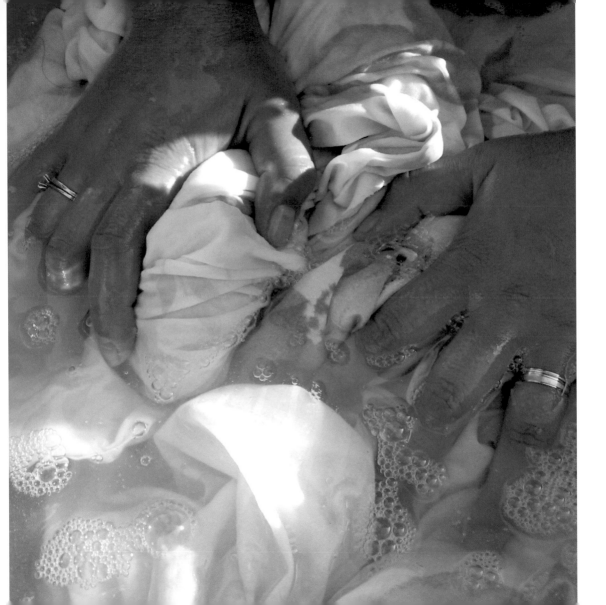

Odour Removal

Smoke Gets in Your Clothes

To remove smoke smells from clothes, whether from cigarettes or bonfires, add half a cup of bicarbonate of soda at the rinse cycle. This will also get rid of the smell of mothballs or musty cupboards that might be lingering on your clothes.

Alternatively, soak smoky clothes in a solution of bicarb before you put them in the washing machine. This works well for clothes tainted with smoke, or clothes that have not been dried properly and have developed that horribly rancid 'old dishcloth' smell.

Perspiration Inspiration

Bicarbonate of soda not only deals with sweat stains but banishes the smell too. This can be a real boon as work clothes and sports garb can still retain that acrid smell (sorry, there's no nicer way to describe it) even after normal laundering. Soak offending items overnight in a solution of 100 g/3½ oz/½ cup bicarb and 4 1/7 pts/4 qts warm water before laundering as usual.

Damp Patches

If there are areas of mildew on clothing that has been in storage, rub with a paste of bicarb and water before washing.

Bin and Gone

Banish the whiff of fusty laundry bins by sprinkling bicarb over the dirty washing. This also prevents the bin itself from becoming smelly. Add a further handful of bicarb each time you put another load of grubby clothing in.

Feeling Creative?

Popping a home-made sachet into your laundry bin will absorb unpleasant smells for months. Simply cut out a circle of lightweight fabric, about the size of a dinner plate, and place a large handful of bicarb in the centre. Add some lavender or a few drops of your favourite essential oil. Gather the fabric up around the powder and secure with an elastic band. Tie a ribbon around the band to hang against the side of the bin or drop it into the bottom. Nice.

Sick Note

Sorry to mention it again, but the smell of sick has a nasty habit of lingering on clothing, even after laundering, so scrape off as much as possible, run the affected area of the garment under the tap, then sprinkle the area liberally with bicarb and leave to act for about an hour. Scrub and rinse, then launder as normal.

Dunk the Skunk

OK, it's not likely to happen, but if you ever get squirted by a skunk, you will be tempted to put all your clothes in the bin – and preferably one several miles from home. Stop: you may be able to rescue them if you act quickly. Try soaking the stinky clothes overnight in a solution of 100 g/3½ oz/½ cup bicarbonate of soda to 4 ½ pts/4 qts water, repeat if necessary, then wash as normal.

Give Your Washing Machine a Holiday

If you find your washing machine smells less than fresh when you get back from your holidays, sprinkle a little bicarb and leave the door open overnight. Next morning, clean the inside and outside of your washing machine and tumble drier with bicarb sprinkled on a damp cloth or sponge.

Next time you go away, sprinkle in the bicarb before you go. No need to rinse out, just leave it in there for the next wash.

Condition & Colour

Getting clothes clean is important but so is the condition and colour of your laundry when it comes out of the machine (or the sink, if you are hand washing). Try these simple ways to bring your laundry right up to scratch, and you will have softer, brighter, more vibrant clothes in a jiffy.

Soft & Bright

Soften It Up

Bicarb not only washes clothes, it also works well as a fabric softener. Its gentle action is ideal if you find commercial fabric conditioners irritate your skin, or you want to avoid all that plastic packaging. Try adding 100 g/3½ oz/½ cup bicarb at the rinse stage or mix with a little warm water and pour into the conditioner drawer of your washing machine. It leaves clothes smelling fresh, but without the artificial perfumed smell of shop-bought softeners. Bicarb as a softener is particularly good for keeping towels fluffy and soft.

Alternatively, mix two parts water, one part bicarbonate of soda and one part white vinegar. It will fizz up like mad, but once it has calmed down completely you can pour it into a plastic bottle for storage. Be sure to label it clearly and keep it out of the reach of children. Add about 50 ml/2 fl oz/¼ cup of the mixture to the fabric softener drawer in your machine per wash.

Whiter Than White

If you habitually add bleach to the detergent in your washing machine, also add 50 g/2 oz/¼ cup bicarbonate of soda if you are using a front-loading machine, or twice as much for top-loaders. Your usual bleach will not only work harder but you will only need to use half the quantity that you normally do. Your white wash will come up brighter, and bicarb is much kinder to your clothes. It is gentle on coloured items too. Alternatively, add one part bicarb and one part freshly squeezed lemon juice to your whites wash to get clothes really clean.

Personal Care

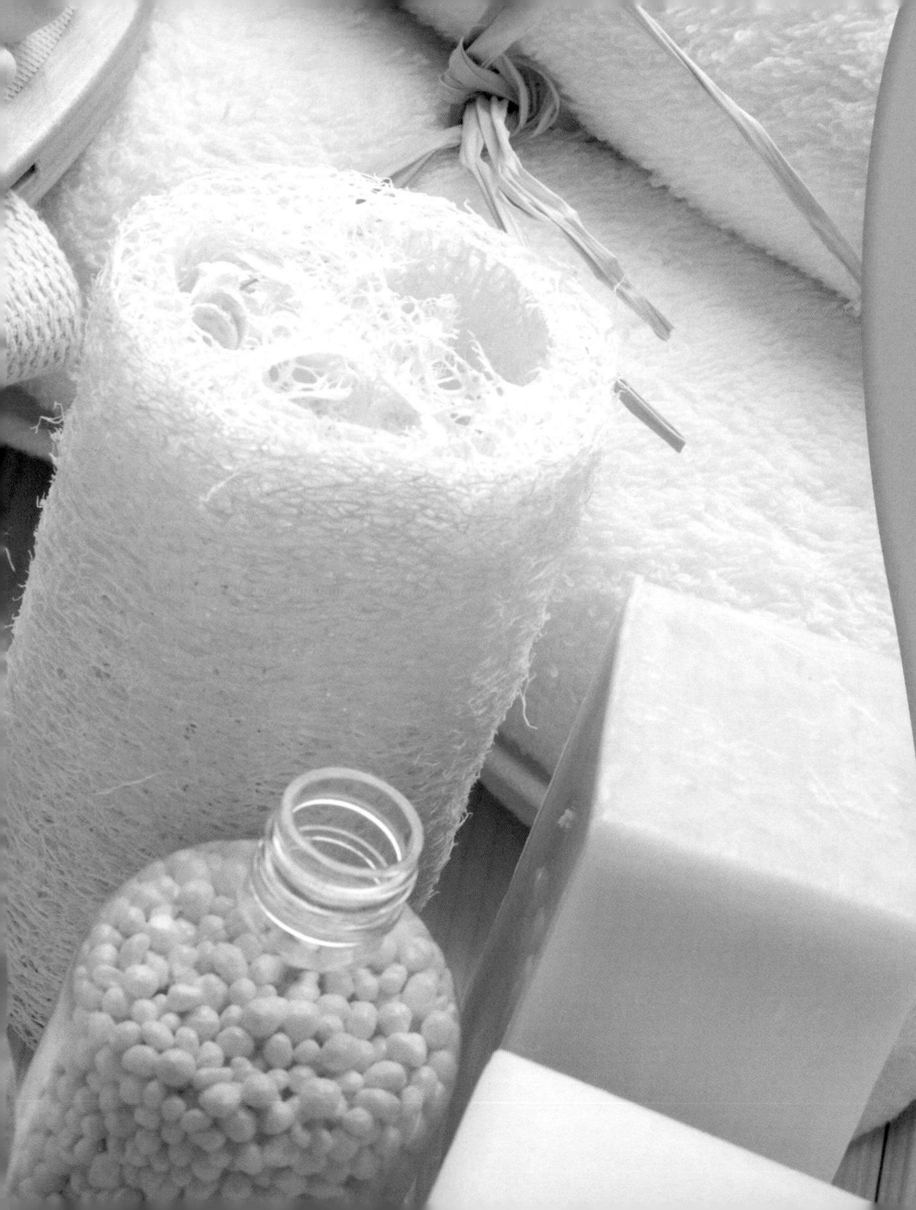

Appearance

It is amazing that something as inexpensive as bicarbonate of soda can be a boon to everyone who cares about appearance, without having to pay a fortune. Bicarbonate of soda can be used in so many different ways to improve appearance, from alleviating irritating little rashes to making your skin so healthy-looking that people will believe you have used one of the luxury brands. Bicarbonate of soda is so versatile that you can use it on most parts of your body.

Skin

Soothe Allergic Skin

If you are allergic to anything at all, from food to medication, more often than not the symptoms will show up on the skin in the form of a rash. These skin allergies can often take the shape of unsightly hives, which are very itchy and take ages to disappear.

Although bicarbonate of soda cannot help with curing the allergy, using it in your bath water can bring enormous relief from the itching. Using 400 g/14 oz/2 cups in a cool bath should be enough to relieve the symptoms, but if you find that particular areas of the body are causing you more intense discomfort, then neat bicarbonate of soda can be rubbed into these areas for even more relief.

Ease Sunburn and Windburn

In much the same way as bicarb brings relief from allergic skin reactions, 400 g/14 oz/2 cups of

bicarbonate of soda can also be added to a cool bath if you are suffering from sunburn or have been out in a strong wind for too long. It will help take the burn out of the skin, but obviously cannot remedy any other symptoms of too much sun that you might be suffering.

If you've been on a boat and got windburn on your face, the bicarb will soon soothe your skin.

 CAUTION: Remember that everyone's skin is different. Please be aware that these tips do not replace the advice of your doctor or beauty professional and, although bicarbonate of soda is mild and natural, particular care should still be taken with sensitive skin. When using any new product, it is advisable to do a patch test before using on large areas.

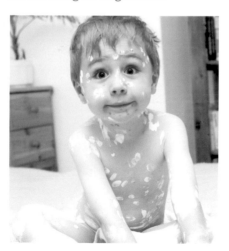

Deal with Chickenpox Spots

If you cast your mind back to when you were a child and had those awful childhood illnesses, you may well remember how uncomfortable chickenpox was. There was the constant itching and the constant rebukes from your parents that you would scar yourself if you scratched the spots.

Well, if you ever come into contact with some poor soul who is suffering from similar problems, suggest that they make a paste of bicarbonate of

soda and a little cool water. This can be applied directly to the spots without any fear of additional irritation. The paste will soothe the itching and make the patient feel much more comfortable.

Relieve Insect Bites and Stings

How annoying is it when those nasty little bugs creep up on you and give you such a bite that you feel as if a lion has mauled you? The worst ones are those mosquitoes, which you can only hear and not see. These insect bites and stings are so itchy or painful that sometimes you cannot sleep for the pain or desire to scratch. And you know it is best not to touch them.

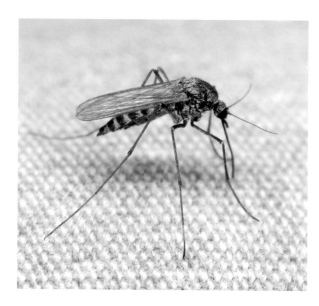

If you mix some bicarbonate of soda into a paste with a little water and then smear it over the bite or sting, you will be amazed at how quickly it gives you relief from the itching.

Help Acne

In a similar way to the relief of the itching of chickenpox spots, bicarbonate of soda can be used to relieve acne, which can also be very itchy. The bicarbonate of soda is again mixed with a little cool, but not cold, water into a paste. The

paste should be applied to the affected area and left to dry for 5–10 minutes. The paste will relieve the irritation of the acne spots and will also act as an extremely mild and non-intrusive facial scrub. It will wash away easily from the face with cool water without damaging the skin in any way.

Ease Razor Burn

When we talk about razor burn, we automatically think of men shaving their face. But you can get razor burn on your legs or under your armpits too and that really can be just as uncomfortable for women as facial razor burn is for men. Once again, a bicarbonate of soda paste can come to the rescue quite quickly. Make the paste quite a thin one and dab it on to the area affected by the razor burn to get almost immediate relief from the burning sensation.

However, if that does not work quite as quickly as you would like, then you could sprinkle neat bicarbonate of soda on to the skin.

Exfoliate

Mixing a paste of three parts bicarbonate of soda to one part water is a brilliant way of exfoliating the skin all over the body. It is also really inexpensive and easy to use.

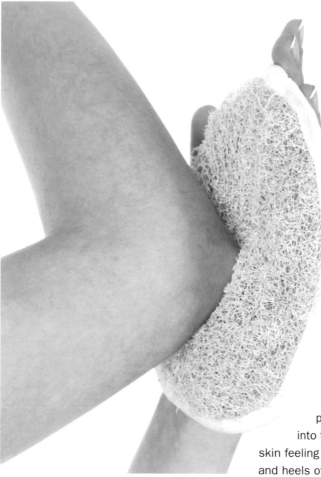

Facial Scrub

The same quantities used for exfoliation can be mixed to a paste and applied to the face as a facial scrub. Use it after washing for that freshly scrubbed feeling. Apply the facial scrub in a circular movement to the face and then rinse it off with cool water.

Rough Patches

Those little parts of your body that seem to acquire hard skin for no particular reason, for example the elbows and knees, will definitely benefit from a bicarbonate of soda paste. Mix to a paste and rub into the affected area to leave the skin feeling softer. It works on the soles and heels of the feet as well.

Pen Stains on Fingertips

You know when an ink, ballpoint or felt-tip pen decides to leak all over your fingers? Well, an easy way to remove this type of stain is to rub your hands together with a mix of bicarbonate of soda and water. Rub away at the stain with confidence that it will not affect your skin but it will remove the stain. It will also remove some paint stains.

Hand Moisturizer

Remember the advertisement that went, 'Now hands that do dishes can feel soft as your face'? Well, that can be true of any washday red hands if you add a little bicarbonate of soda into your washing-up water. The bicarb softens your hands, no matter how long you have to stand at the kitchen sink.

Bicarb Bath Salts

Lying in a hot bath when your body is aching from exertion must be one of the nicest feelings. But you do not have to pay expensive prices to buy salts to add to the bath. Mix equal quantities of sea salt and bicarbonate of soda (about 200 g/7 oz/1 cup is fine for one bath) and 250 ml/8 fl oz/1 cup of the shampoo of your choice while the water is running.

Once the bath is half-full, add your mixture and then top the bath up with more water. You will get some bubbles if you swirl your hand around and then you can soak to your heart's content.

Bicarb Bath Oil

Making bath oil out of bicarbonate of soda is just as easy as making bath salts, but this time you should let the bath fill completely before adding the mixture. Use the following ingredients and mix them in advance of your bath time.

In a mixing bowl, add 225 g/8 oz/1 cup sea salt, 4 tablespoons bicarbonate of soda, 250 ml/8 fl oz/1 cup honey and 450 ml/16 fl oz/1¾ cups milk. Once the bath is full, pour the mixture in and add 115 ml/4 fl oz/½ cup of any of your favourite proprietary baby oil. This will give your skin a lovely soft and silky feel and you are sure to feel nicely relaxed too.

 CAUTION: Do not attempt to keep any over for the next bath – they have to be used fresh.

Dry Skin Bath Additive

If you are suffering with dry skin from the central heating or because of an allergy, you can relieve this by mixing 100 g/3½ oz/½ cup bicarbonate of soda, 75 g/3 oz/1 cup oatmeal, 250 ml/8 fl oz/1 cup warm water and 1 tablespoon vanilla essence (or your favourite) into a paste. Let the mixture run under the tap while you are running your bath. It certainly stops the itching.

Hair

Conditioning the Hair

Sometimes, if you use the same shampoo or mousse all the time, you can get a build-up on your hair, making it feel lank, greasy, drab and lifeless. By adding 2 teaspoons bicarbonate of soda to a teaspoon of your favourite shampoo and mixing them together, you can help to condition your hair and remove the build-up. You just have to wash and then rinse your hair as normal with your mixture. This mix gives your hair a good shine and leaves it feeling really clean.

Hairbrushes and Combs

How embarrassing is it when someone asks you if they can borrow one of your hairbrushes or combs and they look absolutely filthy, or even if they are visible when visitors appear at your home? It cannot be avoided easily because of hair, dirt and grime, but why not try boiling them (if they are of a suitable material) for about 10 minutes in water and bicarbonate of soda? Just add 200 g/7 oz/1 cup bicarb to the water and boil the dirt and grime away.

Teeth & Mouth

Whitening Teeth and Stain Removal

We all know that some of the proprietary brands of whitening toothpastes contain bicarbonate of soda. But you can use a little bicarb once a week on your wet toothbrush to do much the same thing. Just dip your wet toothbrush into a little bicarbonate of soda and brush your teeth as normal. The bicarb also helps remove any stubborn stains from your teeth. You should not swallow this paste and only use it if there are no signs of rawness on your gums.

For badly stained teeth, you could add a little hydrogen peroxide, although this should be used with care and definitely not swallowed.

Plaque Removal

We all know how quickly plaque and tartar build up on the teeth and how much better the teeth

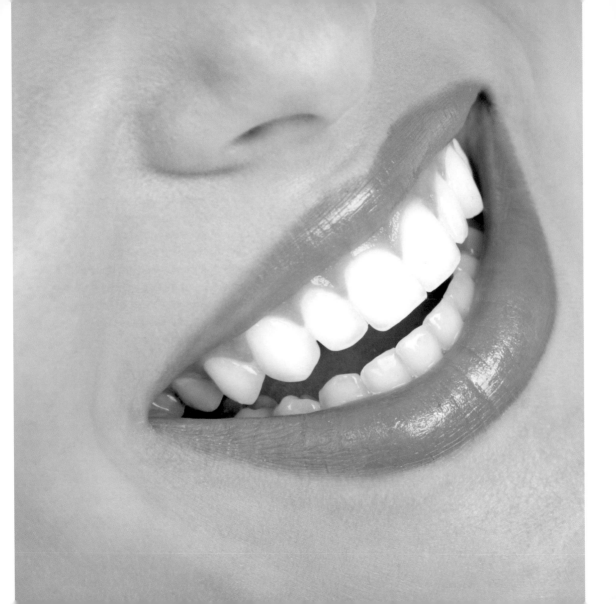

feel when the dentist has cleaned them. Using bicarbonate of soda mixed with iodized salt on a wet toothbrush can help keep the plaque down.

Bespoke Toothpaste

If you like the taste of your regular toothpaste, then why not add some bicarbonate of soda to it? By mixing the two together and placing the mixture into a sealed container, the mix will keep for ages. It will help keep down the plaque, as well as the stains, and keep your teeth whiter at the same time.

Cleaning Toothbrushes

We all do it – we grab the toothbrush without really giving any thought to the germs or bacteria that might be hiding on it. We brush away innocently. But the toothbrush can retain any number of germs and bacteria. To kill them, you can soak

your toothbrush in a weak paste made of bicarbonate of soda and water mixed together. The toothbrush would be fine soaking all night and at least you would have peace of mind that it is clean when you come to use it the next morning.

Breath Freshener

You can make your own breath freshener – particularly useful if you dined on onions or garlic the night before! Using a gargle made of ½ teaspoon bicarbonate of soda mixed with water could be part of your daily routine. This mix can kill off any germs and leave your mouth feeling much fresher.

When you really feel as if you should hide away from the world because of lingering breath odour, use ⅛ teaspoon bicarbonate of soda mixed with water as an instant breath freshener. Swish it around like a mouthwash and it will neutralize the lingering bad odours.

Cleaning on the Move

Wherever you are, if you are expecting to get close or speak to people, it is always nice to feel confident that your mouth is fresh. If you are off to a meeting or a hot date, then dipping a piece of chewing gum into water and then into a small amount of bicarbonate of soda can

really help to make your mouth feel fresh and clean. It might be worth keeping a phial of it in your handbag or in the glove compartment of your car, just in case?

Nasty Cold Sores

Don't you just hate them? One minute you are just fine and then this huge, ugly and unsightly 'thing' appears on your lip. Don't panic though; you can treat a cold sore with some bicarbonate of soda mixed with a little water. Gently – and we mean gently – rub the sore. You will feel relief and it will speed up the healing process.

 CAUTION: The cold sore virus can be carried on your toothbrush, so remember to soak that in bicarb too.

Retainers and Dentures

Braces, retainers and dentures can be soaked overnight in a mixture of bicarbonate of soda and water. This mixture helps to kill off any germs that might have accumulated on them. It also helps to keep retainers looking bright and shiny. If dentures are badly stained, they can be scrubbed with a toothbrush that has been dipped in a bicarbonate of soda paste.

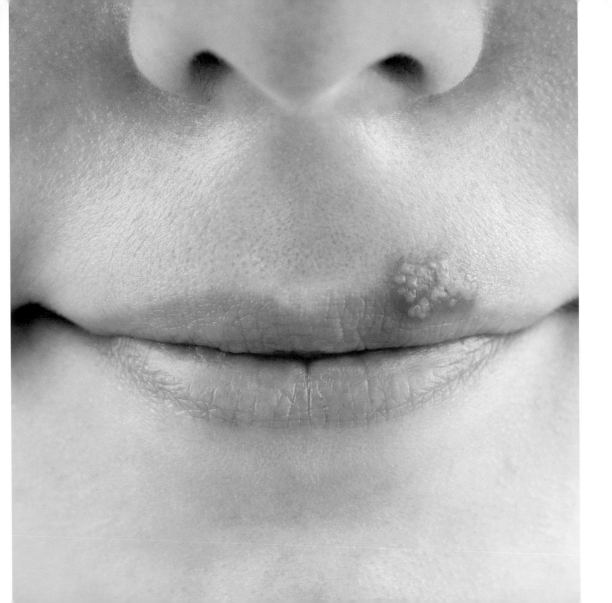

Odour & Feet

Smelly Feet

Smelly or itchy feet can really plague some people in the hot weather, but using bicarbonate of soda as a talcum powder first thing in the morning can help keep the feet dry and itch-free all day. Use it before you put on your socks and footwear. Good for your feet and good for those around them too!

Underarm Deodorant

You can even use bicarbonate of soda to make an underarm deodorant – how cool is that? Just mix together 4 teaspoons potassium alum (which you can buy in block form; it is used as an aftershave and rubbed over a freshly shaven face) with 2 teaspoons bicarbonate of soda and 250 ml/8 fl oz/1 cup alcohol.

You can decant the mixture into a suitable plastic spray-capped bottle and it can be used daily. It keeps you fresh and dry all day long and the added bonus is that it does not contain any harmful chemicals.

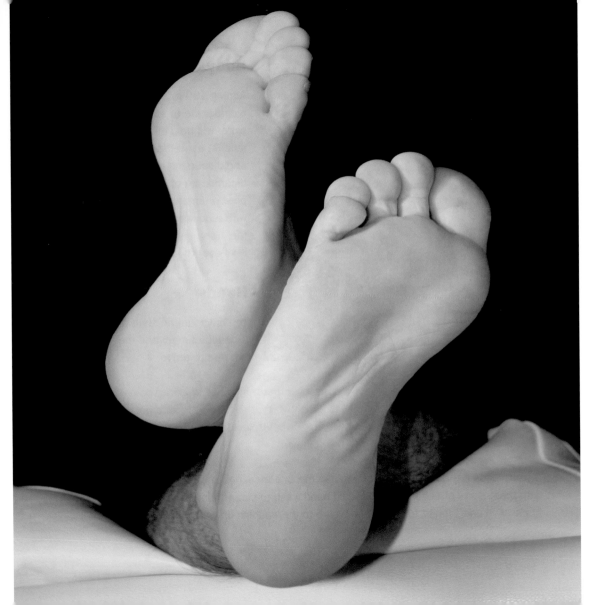

Body Odour

Adding 100 g/3½ oz/½ cup bicarbonate of soda to the bath water will help eliminate perspiration smells, as well as neutralizing acids and dispersing oil. If you use bicarbonate of soda like a talcum powder and apply it to your body with a powder puff, then you will feel fresh all day long. You can use it all over to keep the odours at bay.

Hands

You have just finished preparing a beautiful meal and the guests are due to arrive.

You put your hands to your face, only to discover that they smell of garlic, onion and any combination of different food smells. Don't worry though; a little bicarbonate of soda used as soap with a little water will quickly neutralize the smells and make you feel much better.

If your hands are particularly dirty, you can mix one measure of bicarbonate of soda with three parts of either water or liquid hand wash to clean them up, then rinse off the grime. And don't forget to add some bicarb to the washing-up water, too.

Talcum Powder

Bicarbonate of soda straight from the pack works just like a talcum powder if you sprinkle it on to your skin. The bicarb helps absorb any excess moisture, particularly when the temperature is high, it is a bit too humid and you have been perspiring. Just like your favourite talcum powder, the bicarbonate of soda makes you feel fresh all over.

Tired Feet

If you have been on your feet all day at work, or spent the day 'shopping till you nearly dropped', then a foot bath can feel wonderful. If you mix 200 g/7 oz/1 cup bicarbonate of soda into some comfortably hot water, lower your feet gently in and then relax, you will be amazed how wonderful your feet feel afterwards. And your feet get softened too!

Baby Care

Because bicarbonate of soda is not harmful to children or babies, it is the perfect way to keep the nursery clean and fresh-smelling. Bicarbonate of soda can be used for many baby-related tasks. It can be used to clean baby and his or her toys, as well as to eliminate those sometimes overpowering baby smells around the home. You can use bicarbonate of soda with confidence around baby; it contains nothing that can irritate the skin or cause any allergic reactions.

Bathing & Treatments

Baby Bath

Adding 200 g/7 oz/1 cup bicarbonate of soda to baby's bath can help keep their skin wonderfully soft. You can use it instead of a bubble bath and rest easy that it will not irritate baby's skin.

Cleaning Bath Toys

Baby's bath toys can get a build-up of limescale. To remove this, sprinkle some bicarbonate of soda on to a damp cloth and scrub clean. The bicarb also eliminates any lingering smells on the toys. For really stubborn limescale on the toys, you could make a paste of bicarb and water and scrub with an old toothbrush.

Cradle Cap Treatment

To effectively remove what can be unsightly cradle cap, mix together some bicarbonate of soda and water. Smear it on to the affected area about an hour before bath time, then rinse the mixture off in the bath. Adding some bicarbonate of soda to olive oil is another cradle cap treatment. Mix the two together and apply to the affected area by smearing it on.

 CAUTION: It is quite messy, so use old sheets if you can.

Nappy Rash Treatment

Nappy rash can be so uncomfortable for baby. If you have tried all the proprietary brands and they have not worked, try putting 1–2 teaspoons bicarbonate of soda in the bath water after you've washed baby's face. The bicarb soothes baby's bottom and helps prevent nappy rash coming back.

If the nappy rash is causing serious discomfort, you can use bicarbonate of soda at every nappy change. Dissolve 1 teaspoon in a cup of warm water and apply it to baby's bottom. Apply the mixture around the whole of the affected area and allow the mixture to air-dry naturally.

Thrush in the Mouth Treatment

If baby is suffering from thrush in the mouth, which is quite common in infants, mix together ¼ teaspoon bicarb, 250 ml/8 fl oz/1 cup water and a drop of liquid soap. Using a spotlessly clean cotton cloth, dab the mixture on to baby's tongue 4–6 times a day. It will not hurt baby.

Cleaning & Odour Removal

Cot and Playpen

Because bicarbonate of soda is not harmful to children or babies, it can be used to freshen up their room. It deodorizes smells and leaves the whole room smelling fresh.

Baby's cot can get quite grimy, particularly when they start pulling themselves up. To get rid of finger marks and grease, use a mixture of 100 g/3½ oz/½ cup bicarb, 50 ml/2 fl oz/¼ cup white vinegar and 2.25 l/4 pints/2¼ qts warm water. This mixture will get rid of smells in the cot too.

Any urine smells lingering in the nursery or child's room can be eliminated by sprinkling some bicarbonate of soda on to the mattress of the cot or the bed when changing the sheets.

The playpen can get just as mucky, if not more so, than the cot – it often has lingering food, vomit and urine smells. Use the same mixture as for cleaning the cot. This will deodorize the playpen and make it a nicer place for baby to play.

Bottles

Put a couple of tablespoons of bicarbonate of soda in baby's bottles and fill them with water. Leave them overnight to soak and the sour smells will disappear.

If your baby's bottles can be put in the dishwasher, add some bicarbonate of soda to the powder or tablet area. This will ensure that the dishwasher is clean enough to wash baby's bottles.

Teats and Dummies

When teats and dummies are not in use, they can be kept in cold water to which a pinch of bicarbonate of soda has been added.

 CAUTION: Rinse well afterwards.

Teats and dummies can get discoloured and stained, particularly when baby has moved on to solids. To remove the stains, mix a weak paste of bicarbonate of soda and water, and give them a good scrub.

Toy Box

You can clean baby's toy box too. Once a month, mix together 200 g/7 oz/1 cup bicarbonate of soda with an equal quantity of white vinegar. Use the mixture to clean the toy box and wipe all toys made of suitable material.

Cuddly Toys

Bicarb works on soft cuddly toys just as well. Put them in a clean pillowcase and add some bicarbonate of soda, which you should sprinkle over the toys. If you give them a good shake they will come out smelling as fresh as daisies.

Crayon Marks

It is going to happen as your baby starts to get mobile – crayon on the walls and paintwork. Mix up a paste of bicarbonate of soda and water. With an old toothbrush, apply the paste to the mark – magic, it's gone!

Washing Cloth Nappies

Some babies can get an allergy to washing powder that can cause nappy rash. Try putting 200 g/7 oz/1 cup bicarbonate of soda in the wash to help avoid this.

The washing powder you use regularly can build up on nappies and baby clothes over time. To eliminate the build-up of the washing powder, replace it for one wash with 200 g/7 oz/1 cup bicarbonate of soda. Use similar quantities to your washing powder. It won't lather so much but it will reduce the build-up.

You can keep your nappies looking bright and white if you add 100 g/3½ oz/½ cup bicarb to your washing machine. Put it in the powder chamber of your machine and you will be proud to hang those white nappies out to dry!

Nappy Bucket

If you use terry or cloth nappies, the bucket will inevitably get smelly. Sprinkle some bicarbonate of soda (about 200 g/7 oz/1 cup) over each nappy as you add it to the bucket. This will deodorize the smells until washday.

Deodorizing Clothes

To eliminate baby smells from their clothes, and from your own, add 100 g/3½ oz/½ cup bicarbonate of soda to water and soak the clothes for a few hours before washing.

If baby's clothes don't need a wash, but just need to be freshened up, you can sprinkle them with some bicarbonate of soda to deodorize any lingering odours. Leave them overnight if you can and the stale smells will be gone by the morning.

Health

Bicarbonate of soda is so safe to use that you can also ingest it. Although it is artificially produced for mass use, it is a compound that also occurs naturally as sedimentary mineral deposits; it is not the usual horrible synthetic chemicals that we often wash down the drain, so it is not harmful to us or to the environment as a whole. That's why it is so good to use as a healthcare product. The other added advantage is that it is cheap to buy and readily available in varying quantities.

Feel Better

Sports Drink

A really inexpensive and effective sports drink for all the family can be made by adding bicarbonate of soda to a sugar-free children's drink. Mix 1 teaspoon bicarbonate of soda, 1 tablespoon salt, 5 teaspoons sugar and 2.25 l/4 pints/2¼ qts water with the sugar-free drink and refrigerate it. It keeps for quite a while.

Your bicarb sports drink can be used in exactly the same way as any proprietary sports drink. Decant it into a reusable bottle and it can go anywhere with you or the children. It must be better for them than those fizzy, sugary drinks you can buy and it's cheaper!

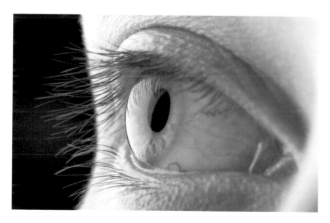

Itchy Eyes

If you are suffering from itchy eyes, which could be the result of an allergy to pollen, you can soothe them with a bicarbonate of soda eyewash. Add ½ teaspoon bicarbonate of soda to half a glass of water and bathe the eyes with an eyebath. The mixture will keep in the fridge for a couple of days.

Contact Lenses

Daily-wear contact lenses need cleaning every day. Add a pinch of bicarbonate of soda to your cleaner to make it mildly abrasive. This helps prevent build-up and could extend the life of the lenses.

Reaction to Poison Ivy

Ever touched poison ivy by mistake and wished you never had? Really painful, isn't it? You can deal with the resultant rash by soaking in a bath containing 100 g/3½ oz/½ cup bicarbonate of soda.

 CAUTION: Remember that the tips given here are not a substitute for a medical professional's assessment.

Stuffy Nose

All bunged up? Using bicarbonate of soda in a vaporizer can really help to clear a stuffy nose quickly. Fill the vaporizer with water to the indicated level, then add and stir in 1 teaspoon salt and 1 teaspoon bicarb until they are dissolved. Now breathe in the vapour to clear the congestion.

This is a really good tip for those who cannot or do not wish to use proprietary decongestants. Pregnant women, children and those who have medical conditions are often prevented from using medication. Bicarbonate of soda is a safe alternative for everyone, whatever their circumstances.

Not only does bicarbonate of soda in your vaporizer help clear your stuffy nose and reduce the symptoms of a cold, but it also deodorizes your house and removes lingering odours. How's that for good value?

Fever Reduction

If one of your loved ones has a high fever, there's nothing quite as effective and speedy to reduce the temperature as a bicarbonate of soda bath. Add some bicarb to a cool, but not cold, bath and the temperature will plummet fast.

Children's high temperatures will really drop quickly if you pop them into the cool bicarbonate of soda bath. They will feel much more comfortable. Of course, the added bonus is that it washes them squeaky clean at the same time as lowering their temperature.

Indigestion

Bicarbonate of soda is known to be an effective antacid, so it can work for most stomach discomforts, particularly things such as indigestion and heartburn, which can make you feel really uncomfortable.

If you are one of those unfortunate people who regularly suffer from that bloated feeling after a meal, try sprinkling a little bicarbonate of soda on to your food. It helps to reduce the build-up of gas that causes the bloated feeling. You can use bicarb whenever you have excessive bloating for whatever reason.

Some foods and drinks can give you heartburn every time you eat or drink them. But you don't have to suffer in silence or give up consuming them. Add 1 teaspoon bicarbonate of soda to a glass of water and the burning feeling will disappear really quickly.

Other types of food, perhaps those that are greasy or really rich, can give you an upset stomach. This is particularly true if you don't make a habit of eating a good old fry-up. When you do, try adding 1 teaspoon bicarbonate of soda to your glass of water and the churning in the stomach soon settles down.

Outside &
Maintenance

Garden & DIY

Probably the most fundamental reason for using bicarbonate
of soda in the garden, greenhouse and the outside of your house
is the cost. Bicarb is so cheap that needing large quantities of it to
complete the job is not a problem. In addition, because bicarb is a natural
product, the fact that you are using large amounts outside does not
pose a threat to the environment. There are so many useful
outside jobs you can do with it that you'll be amazed.

Tools & Equipment

Preserve those Tools

Shovels, forks, rakes – indeed all your garden tools – should be put away for the winter in a clean state to help preserve them. You can make a thick paste with bicarb and white vinegar to give them a good clean. This will also stop rust forming over the winter months.

Polished Pots

If your garden pots are not frost-resistant, then they should also be stored away before winter sets in. But if you put them away dirty, you will find yourself faced with twice the job of cleaning them in the spring. Use the bicarbonate of soda and white vinegar paste to clean them and then wash this off with some hot soapy water. They will sparkle in the spring!

Potato Trick

If any of your garden tools have been neglected and become rusty, you can use a potato and some bicarbonate of soda to renew them. Just peel the potato and keep dipping it in the bicarbonate of soda and rub until the rust has all been removed from the metal tool. Shiny again!

Happy Pots

If you line your terracotta pots with a thin layer of bicarbonate of soda before adding the soil, it will help to keep the soil fresh. It will kill any lingering bacteria and germs that may have remained from the last plants.

Outdoor Furniture & Surfaces

Wash that Plastic

Garden furniture is not cheap, so if you want to prolong its life, you have to take care of it. Plastic garden furniture should be washed before it is put into storage for the winter. Add 200 g/7 oz/1 cup bicarbonate of soda to some hot soapy water and wash the furniture with the mix. This will help remove stains too.

Bright Whites

If your garden furniture is all white in colour, you can keep it that way by adding some lemon juice to your bicarbonate of soda mix. Squeeze the lemon juice into your hot water and bicarbonate of soda mix. It is as good as bleaching the white furniture.

Clean the Barbecue

Cleaning the barbecue can be a tiresome job and quite a messy one. To clean the appliance without causing any scratching to the stainless steel surface, try sprinkling some bicarbonate of soda straight on to a damp brush. Coat the barbecue well with the bicarbonate of soda, give it a good scrub, then rinse clean.

Secure Repairs

Do you have any repairs that need doing on your garden furniture? If the furniture is plastic and you intend to use superglue, you can make the repair even stronger by adding a little bicarb to the glue. Sprinkle the bicarbonate of soda on to the superglue while it is still wet.

Pool Perfection

Whilst you are busy cleaning the garden in general, the children's paddling pool might benefit from a bit of tender loving care too. Get rid of any mould or mildew by rubbing it inside and out with a mix of water, bicarbonate of soda and white vinegar. Good as new and safe for the children next year.

Birdbaths

Even the birdbath needs a good clean from time to time, but don't worry about harming the creatures that drink from it or bathe in it. Using bicarbonate of soda and water mixed to a paste will remove all stubborn grime and will not affect the birds at all.

Pavements, Patios and Decking

Although it might take a whole box of bicarbonate of soda to clean your pavements and patio areas, it is still effective and inexpensive. Wet the bicarb a little bit after you've sprinkled it over your concrete areas and then scrub with a stiff garden broom to remove stains and moss from the concrete.

You could use the sprinkling and brushing technique for cleaning your wooden decking too. Sprinkle the bicarb on and brush with a dry broom, but wash away the dirt and grime with water afterwards.

Hygienic Hammocks

Garden hammocks can be wiped down with a bicarbonate of soda and water mix. This will get rid of any built-up grime and also bring out their colour, not to mention making them smell really fresh for you too when you swing next summer.

Cleaner and Safer Toys

Children's garden toys, such as trampolines or slides and seesaws, as well as plastic cars and scooters, all need the bicarbonate of soda touch. Bicarb mixed with water will bring the colours back to life and keep the dust levels at a minimum. It must be better and healthier for the children to use clean toys in the garden?

Plastic garden toys can become brittle if they are stored away in a warm environment during the winter months. To counteract this effect, you can wipe the toys with a mixture of bicarbonate of soda and white vinegar. The two ingredients when mixed together will clean and protect the toys at the same time.

Snow and Ice

Bicarbonate of soda acts very much like salt on snow and ice. Because the bicarb, like salt, freezes at a lower temperature than water, it is an ideal way to melt snow once it has lain on your concrete. Sprinkle it over freshly fallen snow to help it melt quickly, then add some more in case the slush freezes.

For a more preventative measure, if snow and ice are forecast, you can stop the water from freezing overnight by sprinkling bicarbonate of soda on to your concrete. Use it on your pathways and steps in the garden to stop falls and accidents. Sprinkle it generously and cover as much of the concrete as you can.

Greenhouses & Plants

Wash Down

Some people use Jeyes fluid to clean their greenhouses in the autumn. However, if you want an ingredient that is just as effective, but cheaper, you can use bicarbonate of soda. Wash all the surfaces, including the roof and walls, with bicarb and water. It will kill any mould and bacteria that may have formed during the growing season, as well as making the glass shine.

For an alternative mixture to wipe down the surfaces in your greenhouse and remove dirt and grime, mix 4 teaspoons bicarbonate of soda with 120 ml/4 fl oz/½ cup white vinegar and 2.25 l/4 pts/2¼ qts water. You can even put this mixture into a spray bottle, bit by bit, to make it easier to apply. Simply wipe off the dirt with a damp cloth.

Greenhouse Fungicide

Bicarbonate of soda has been used

for years as a mild fungicide, particularly by organic growers. It can be used against powdery mildew and other plant diseases in the greenhouse too. Mix 5 tablespoons bicarbonate of soda with 4 1/7 pts/4 qts water and spray the greenhouse and the affected plants.

Routine for Roses

No matter how careful a gardener you are, mould and mildew still spread really quickly once they take hold. To keep mould on plants, particularly roses, at bay, you can spray with a mixture of 200 g/7 oz/1 cup bicarbonate of soda and 2.25 l/4 pts/2¼ qts water. It really stops the mould from spreading to other plants.

Clean the Bottle

When you have sprayed your garden plants, be sure to kill any mildew or fungus inside the spray nozzle. Add a little bicarbonate of soda and this should do the job for you. Also, before you put your sprayer into storage, do the same thing so it's fresh when watering time comes around again.

 CAUTION: You will have to fill the spray several times, but do make sure you cover all the diseased parts of the plant, including the foliage.

Defeat Disease

You can mix 4 teaspoons bicarbonate of soda into 7.5 l/11¾ pts/7½ qts water to treat diseased plants. Apply the mixture generously to the diseased plants with a hand-held spray.

Free the Lawn from Mould

If your lawn needs some serious attention, there is no better way to give it a boost than bicarbonate of soda, particularly if you have problems with mould. Use 1 tablespoon to 4 l/7 pts/ 4 qts water and spray the lawn, or you can apply it with a watering can. It kills off mould or mildew without having to use a fungicide and gives the lawn a bit of a tonic.

Another way of spreading the bicarbonate of soda on to your lawn is to fill a sock or an old pair of tights with some neat bicarbonate of soda. Beat the side of the sock or the tights

with your hands to release the dust. You get more of an even spread this way and you can always water it in with a hose if it doesn't rain.

Acid or Alkaline?

You can test the acidity of your garden soil by adding a pinch of bicarbonate of soda to 1 tablespoon of your soil. If the mixture fizzes, the soil's pH is probably low and could do with a weak solution of bicarbonate of soda and water. This mixture will help plants such as geraniums, begonias and hydrangeas to bloom. They like a more alkaline soil.

Sweeter Tomatoes

Tomato plants like bicarbonate of soda too! Sprinkle a little around their roots regularly from the first planting. The bicarbonate of soda will reduce their acidity and make them even sweeter to eat.

Blooming Marvellous

After cutting flowers from the garden to take into the house, why not try dipping their stems in a mild bicarbonate of soda and water mix? It will help the flowers to stay fresh longer and kills any bacteria on the stems.

Pest Control

Ants No More!

Ants – whoever needs them? Although they don't do much damage, they are still off-putting, particularly if they are near your doors and windows. You can sprinkle neat bicarbonate of soda around their entrance and exit areas to help get rid of them and hopefully stop them getting into the house.

Ants also seem to love children's sandpits. If they are invading your child's sandpit, then mix a whole box of bicarbonate of soda well into the sand. That should keep them away!

Fleeing Fleas

Fleas do not like bicarbonate of soda either. You can sprinkle it on your lawn, where they often congregate, believe it or not, or you can sprinkle the bicarbonate of soda directly on to your pet to make sure they don't stay on too long.

Clear out Cockroaches

Cockroaches seem to like damp places and places where water congregates. You do not want cockroaches in your home because they spread so many germs. Try to keep damp areas free of the little devils, particularly around water pipes and in cellars, by sprinkling neat bicarbonate of soda in these areas.

Repel Rabbits

No matter how sweet we think they are, rabbits can do a lot of damage in the garden. They can be particularly harmful to plants that have been recently planted and especially to lettuces and cabbages. To ward off rabbits, sprinkle some bicarbonate of soda generously around the vegetable patch.

Say No to Slugs and Snails

Slugs and snails can do even more damage than rabbits. They can destroy a newly planted vegetable or bedding plant overnight by eating the sweet leaves. At least if you use the bicarbonate of soda generously, you should be able to deter most of these pests in one fell swoop, and your plants might actually feed you.

Painting & Decorating

Soften Paintbrushes

If you have been really bad and not cleaned your
paintbrushes properly the last time you
used them, there is an
overnight remedy. Mix
together 200 g/7 oz/1 cup
bicarbonate of soda, 50
g/2 oz/¼ cup white vinegar
and 2.25 l/4 pts/2¼ qts hot
water. Soak the brushes in the
mixture overnight and by morning
they should be soft and ready to use again.

Sticky Residue Remover

When you have painted a window, to remove any glue left from
masking tape, simply make a paste of bicarbonate of soda and water. Rub gently with the
paste to remove the glue residue.

You can use the same mix of bicarbonate of soda and water to remove poster glue from
windows. If you have been advertising a local event and there is some glue residue on the
window, use this mix to remove it completely and easily.

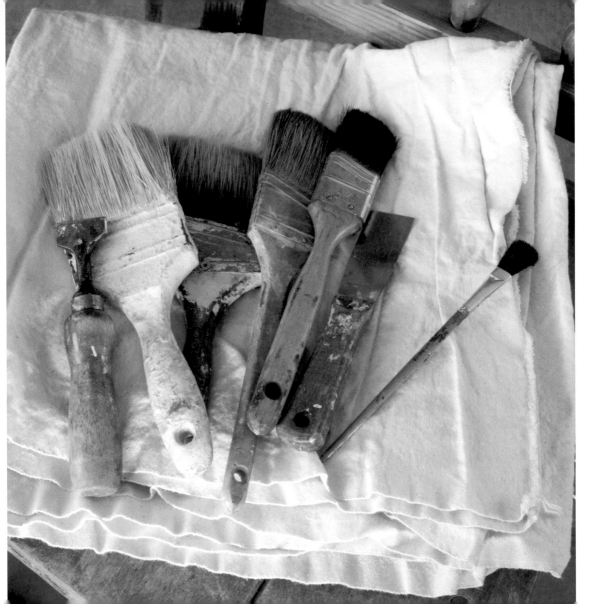

Wash Down Walls

To keep your outside walls clean and bright, you can wash them down with a mixture of 100 g/3½ oz/ ½ cup bicarbonate of soda to each 4 l/7 pts/4 qts water. It will clean efficiently and bring colours back to almost new. The mix can be used on most outside wall materials.

You can use the same mixture (100 g/3½ oz/½ cup bicarbonate of soda to each 4 l/7 pts/4 qts water) to wash down your walls and paintwork before you begin to repaint them. This will remove all the dirt and grime, as well as any of the old paint that may be flaking.

Stripping Wallpaper

If you are planning to strip wallpaper, you can make the job much easier if you add 1 tablespoon bicarbonate of soda to the water. Stir it in until it has dissolved and then wet the walls ready for stripping in the normal way. Let it soak in well and you will be amazed how much more easily the paper comes off.

Vehicles

Cars, bikes and motorbikes can be messy. Let's face it, we spend so much time in our vehicles that it is not surprising that sometimes they are our pride and joy and at others they are the bane of our lives. If we are not worrying about cleaning them then we are worrying about selling them. Keeping them in tiptop condition can only be good when it comes to getting a good price for 'her'. Being safe when we are working on our cars and bikes is so vitally important. Bicarbonate of soda can help with all these things.

Cars & Bikes

Oil and Petrol Spills

If you have a car that leaks oil or petrol on the floor of your garage, then you might like to try the bicarbonate of soda remedy for removing the stain. You just need to sprinkle a mixture of bicarbonate of soda and salt over the spill. Leave it for a while to soak up all the oil or petrol, then all you need to do is sweep the floor.

Even if the oil or petrol leaks on to your driveway, bicarbonate of soda and salt will still lift the stain efficiently.

 CAUTION: Give the mix plenty of time to soak up all the liquid up before you sweep the drive.

Odour Removal

If the car has not been used for some time, it can begin to smell musty inside. Neutralize these smells by sprinkling bicarbonate of soda all over. You can put it in the boot, on the seats and on the mats without any fear of damaging the interior upholstery.

Got smelly ashtrays in your car? You can get rid of that lingering smell by putting 120 ml/4 fl oz/½ cup bicarbonate of soda into the ashtray when it is clean. The bicarbonate of soda will neutralize the smell of the smoke and leave the whole car deodorized and smelling fresher.

Extinguishing Fires

Small fires can be extinguished in a garage or workshop using bicarbonate of soda. Pour a complete box of the bicarb on to the fire and it will extinguish it. It is a good idea to keep some on the shelf just in case, particularly if you regularly use a naked flame in the garage or workshop.

You could also keep some bicarbonate of soda in the boot of your car. It is particularly effective at extinguishing electrical fires. The bicarbonate of soda smothers the fire and brings its temperature down quite quickly.

 CAUTION: Bicarbonate of soda should not be applied to fires in deep fryers as it may cause the grease to splatter.

Bugs on Bikes

To remove bugs that may have got stuck to the windows or wing mirrors of your car or motorbike, you can wash them with a mixture of bicarbonate of soda and water. The bugs will come off quite easily, but you won't scratch the glass.

Rusty Bikes

Bicycles can go rusty quite quickly, particularly if they are housed in a garage or shed. Vinegar and bicarbonate of soda mixed to a paste will give them a general clean, but if there are any rust patches developing, give them a particularly good scrub. The mix should remove the rust quite easily.

Soda Shine

To give the chrome on a bicycle or motorbike an extra shine, apply a bicarbonate of soda and water paste. Smear it on to the chrome and then leave it to dry naturally in the air. When it is dry, you can buff the chrome with a soft cloth and you will be able to see your face in it.

Kids' Stuff

Fun Projects for Children

What a really fun and inexpensive play material bicarbonate of soda can be. Your children will be safely amused for hours making clay, volcanoes and rockets out of bicarbonate of soda. The ingredients are all cheap and easy to get hold of and the added bonus is that there is nothing that can harm the children. What more could a busy parent ask, particularly when the children are also learning useful science- and maths-related stuff at the same time?

Clay

Your children will love making this easy-to-mould clay. It is safe to use and the ingredients are so cheap – certainly much less than most proprietary brands of clay. The children can add food colouring to it if they wish at the very beginning of the project, or they can also decorate it with paint, felt-tip pens or watercolour pens after they have formed their shapes. This recipe makes a good-sized piece of bicarbonate of soda clay.

You Will Need:

1 large box (400 g/14 oz/2 cups) bicarbonate of soda
200 g/7 oz/1 cup cornflour
250 ml/8 fl oz/1 cup water
1 tbsp vegetable oil

1 Place all the ingredients into an old saucepan. You can add the food colouring of your children's choice at this stage, or you can wait and colour at the end.

2 Heat the mixture very slowly, stirring all the time, until the mixture becomes thick and uniform in shape. The heating stage usually takes about 15 minutes or so on a low heat.

3 When the mixture is the right consistency, pour it out of the saucepan and on to an old plate.

4 Cover the clay with a damp cloth and leave it to cool for about an hour.

5 Once the clay has cooled enough to be handled, you can sprinkle some more cornflour on to a wooden board and knead the clay, as you would bread, until the clay is smooth.

6 Now you can let the children loose. The clay can be moulded into any shape that your children wish. They can also use their favourite colours to highlight the shapes. Felt-tip pens or watercolour pens are good for this important task.

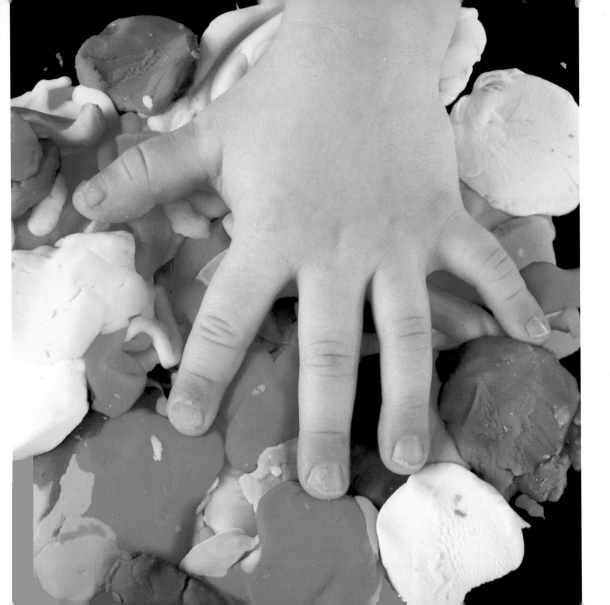

Fizzy Sherbet

There is no cooking involved in making this fun fizzy treat for the children, but getting the taste right can be a case of trial and error. The sugar may have to be adjusted to give the right level of sweetness, which varies from person to person, and the same applies to the tartness or lemony-ness of the sherbet.

Science Made Fun

Making fizzy sherbet is a really good science-related project for the children. They will learn when you explain to them that the sugar and the citric acid give the sherbet the flavour.

You can also talk about carbon dioxide when you see that the citric acid and bicarbonate of soda mixed together give the fizzing action on the tongue. Remember that you only get the fizzing when the sherbet mixes with the saliva on the tongue.

Start With:

2 tsp caster sugar or icing sugar

1 tsp powdered citric acid

½ tsp bicarbonate of soda

1   Mix the ingredients together.
2   Test for sweetness and if necessary add some more sugar.
3   If the sherbet does not taste quite lemony enough, add some more citric acid.
4   If the sherbet isn't quite fizzy enough on your tongue, then add some more bicarbonate of soda.

You can also vary the flavour of your sherbet when you have this initial ingredient mix just right. But remember that all the ingredients have to be dry, otherwise the sherbet will fizz with any liquid you add before it gets on to your tongue. You could try adding vanilla, coffee, cloves, cinnamon, dry ginger and a host of other dry ingredients.

There are various ways you can eat the sherbet – such as simply spooning it into your mouth with a spoon or shovel-shaped straw – but one of the most enjoyable for children is to dip a lollipop or stick of liquorice into the sherbet.

Honeycomb Toffee

Honeycomb toffee, also known by some as 'cinder toffee', is a really good fun project to make with the children. It is also good to eat as a special treat. Honeycomb toffee is a flexible recipe, which means you can add ingredients to it to bring about any number of changes in taste.

You Will Need:

a little butter
1 tsp bicarbonate of soda
150 g/5 oz/¾ cup sugar
4 tbsp honey or golden syrup

1   Lightly grease a baking tray with butter. Melt together the sugar and the honey or golden syrup over a low heat, stirring continuously to prevent burning, until they have caramelized.

2   Slowly add the teaspoonful of bicarbonate of soda. Keep stirring continuously to prevent burning. As the bicarb hits the hot caramel, it will bubble up very quickly and turn a burnt orange colour.

3 Once the mixture has stopped rising because of the bicarbonate of soda, you can carefully remove it from the heat.

4 Pour the toffee mix on to the greased baking tray and cover the hot mix with clingfilm/plastic wrap.

5 Once the toffee mix has cooled sufficiently, place in the fridge to harden.

6 Cut the crunchy toffee into bite-size squares to serve.

 CAUTION: Do not leave young children unattended around hot appliances.

Variations

Children's favourite flavourings, such as orange or strawberry, can be added to the toffee mix at the same time as the bicarbonate of soda. Alternatively, or in addition to the flavourings, you can dip the set squares of toffee into melted dark, white or milk chocolate and leave to set before serving.

You can even sprinkle chopped nuts or fruit-and-nut mix on the top of the still-wet chocolate if you wish.

Inflate a Balloon

Showing children how to inflate a balloon using bicarbonate of soda and vinegar is a really good science-related activity. The project gives them the chance to learn much about chemical reactions and the nature of solids, liquids and gases.

When the bicarbonate of soda and the vinegar you use in this activity are mixed together in your plastic bottle, they form carbon dioxide. As the carbon dioxide gas is made, pressure builds up inside the bottle so that the gas produced escapes into the balloon and inflates it.

You Will Need:

2 heaped tsp bicarbonate of soda
150 ml/5 fl oz/⅔ cup vinegar
plastic drink bottle that can hold 600 ml/1 pt/2½ cups
balloon
plastic funnel

1 Insert your plastic funnel into the neck of the balloon.
2 Spoon into the funnel your 2 heaped teaspoonfuls bicarbonate of soda. Shake the bicarbonate of soda so that it filters through the funnel and into the balloon.
3 Next, you need to pour your vinegar into the plastic bottle, again using your plastic funnel.
4 Whilst holding the balloon tightly so that none of the bicarbonate of soda falls out, you can now stretch the neck of the balloon over the neck of the plastic bottle.
5 Tip the balloon so that the bicarbonate of soda falls into the vinegar and the two mix.

As the chemicals in the vinegar and the bicarbonate of soda mix in the plastic bottle and give off fumes, your balloon will begin to inflate. Once the balloon has stopped inflating, you can remove it from the plastic bottle and tie the balloon in a knot, as you would after the usual means of inflation.

Volcanic Eruption

This inexpensive project is a craft and science project. It takes time to organize, but is worth the effort; the kids will love it.

To Make the 'Volcano' Dough:

700 g/1½ lb/6 cups flour

350 g/12 oz/2 cups salt

4 tbsp vegetable oil

500 ml/18 fl oz/2 cups water

1   In a large bowl, mix all the dry ingredients together.

2   Add half the water and oil and knead to form a dough-like consistency.

3   Add the remaining water and oil gradually, working into a smooth and firm dough. If the mixture feels too dry, add more water.

To Create the Volcano:

plastic bottle that can hold 2 l/3½ pts/2 qts

2 heaped tsp bicarbonate of soda

150 ml/5 fl oz/⅔ cup vinegar

2 l/3½ pts/2 qts warm water (continued overleaf)

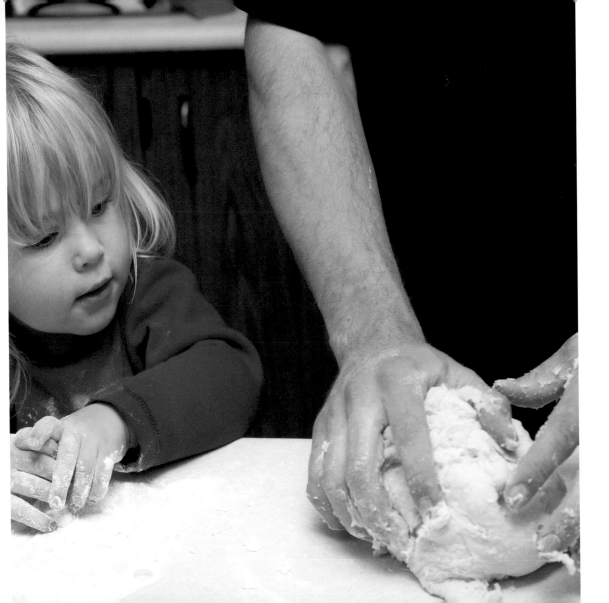

few drops food colouring

few drops liquid soap

2 tbsp bicarbonate of soda

baking tray

1   Stand the clean, dry plastic bottle on your baking tray and shape the dough around the bottle to form a volcano shape. Avoid dropping dough inside the bottle and leave the bottle opening free.

If time allows, let the volcano dry for 2 or 3 days before going any further. You could paint it, providing the bottle opening and the inside of the bottle remain dry. However, if you cannot wait to see the full effect of the volcano, you can make the lava straightaway:

2   Put a few drops of your chosen food colouring into the bottle.
3   Pour over the warm water until you have almost filled the bottle.
4   Add a few drops of liquid soap into the bottle.
5   Then add the bicarb to the soapy water in the bottle.
6   Now to make the lava explode! Just pour in your vinegar.

When the bicarbonate of soda and vinegar mix together, they produce carbon dioxide gas. This bubbles up and forces your coloured lava to erupt.

Mini Rocket

The children will really enjoy this exciting and creative project.

To Make the Rocket Itself:

scissors

plastic canister with lid – an old 35 mm film cartridge case is ideal

sheet A4 paper, coloured if preferred

felt-tip pens or paint (optional)

sticky tape

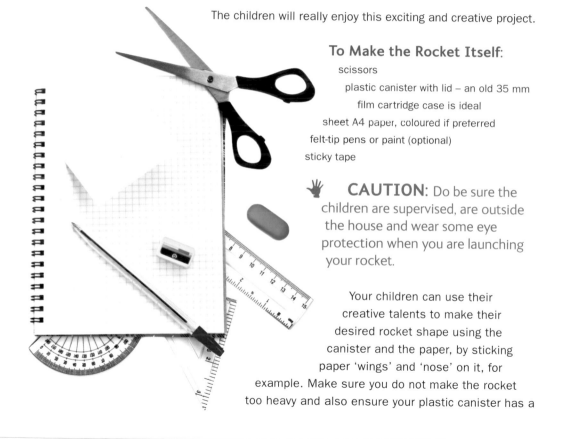

CAUTION: Do be sure the children are supervised, are outside the house and wear some eye protection when you are launching your rocket.

Your children can use their creative talents to make their desired rocket shape using the canister and the paper, by sticking paper 'wings' and 'nose' on it, for example. Make sure you do not make the rocket too heavy and also ensure your plastic canister has a

removable lid. You can make the rocket well in advance of launching it, particularly if the children want to paint or colour their rocket. When you are ready to launch the rocket, it's time to make the rocket fuel.

To Make the Fuel:

½ tsp bicarbonate of soda
½ tsp vinegar
tissue

You should not allow your bicarbonate of soda and vinegar to mix until you are ready to launch the rocket. But when you are ready, you can:

1   Wrap your bicarbonate of soda in a tissue. Make a parcel that will easily fit inside the lid of your plastic canister.
2   Fix the bicarbonate of soda parcel into the lid of the plastic canister with a little sticky tape. Alternatively, make a paste with the bicarbonate of soda and a little water and pack this into the lid.
3   Pour the vinegar into your plastic canister.
4   Carefully put the lid on to the canister, then turn the canister upside down so that the vinegar soaks through the tissue paper and mixes with the bicarbonate of soda.

When the bicarbonate of soda and the vinegar combine, they will make carbon dioxide gas. This gas will force the plastic canister to pop off the lid. The rocket will launch into the air, so stand back and enjoy!

Food

Culinary Tips

In addition to using bicarbonate of soda in recipes, bicarb can also be used in a number of ways to prepare foodstuffs ready for cooking, or to preserve them for future use. Bicarb is a really useful and practical addition to any cupboard shelf. It is cheap and, because it is not harmful to our bodies, you can use it to your heart's content. Bicarb is so versatile that it is amazing it does not carry a huge price tag.

Meat & Poultry

Free of Feathers

Sometimes, when you buy a chicken, there are a few feathers left on the skin. They won't do you any harm, but they can be a bit off-putting. To get rid of these quickly and easily, just rub the skin of the chicken with bicarbonate of soda. Rinse the bicarbonate of soda off before cooking the bird.

Tender Fowl

Bicarbonate of soda is also really good for tenderizing chicken meat. Rub the meat liberally with some bicarbonate of soda and then rinse it thoroughly for a really tender chicken.

Crispy Chicken

If you want the chicken skin to be crispy, you can rub in some bicarbonate of soda before cooking. Crispy skin and tender meat – delicious!

Tone Down Flavour

You can use bicarbonate of soda to reduce the strong flavour of meats such as venison, particularly if you are planning to preserve the meat. Add the bicarbonate of soda to your preserving jar to get rid of an overwhelming flavour.

No Fowl Smell

If you are plucking and preparing your own chicken, you can get rid of the sometimes strong smell by sprinkling some bicarbonate of soda into the cavity before drawing out the giblets.

Pork Crackling

If you like your pork crackling really crisp, then bicarbonate of soda can do this for you. Sprinkle and smear some bicarbonate of soda on to the pork skin before cooking. The bicarb will help the skin to really crisp up for you. Score the skin as usual, with crisscross cuts using a sharp knife. Cutting like this helps to crisp up the skin too and makes it easier to cut and share, because everyone will want some!

Smearing bicarbonate of soda on to the skin of your pork meat will not only give you a crispy crackling, but it will also help keep the meat tender.

Marinade Me Tender

You can add ½ teaspoon bicarbonate of soda to your meat marinade prior to cooking stir-fry dishes. The bicarb will help soften the texture of the meat, particularly if the meat you have bought is one of the tougher cuts. The meat will melt in your mouth.

Vegetables

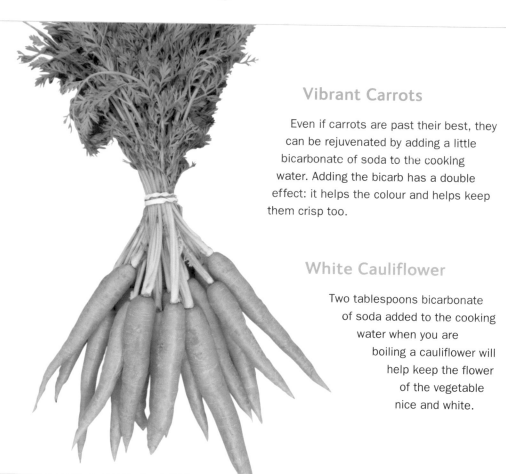

Vibrant Carrots

Even if carrots are past their best, they can be rejuvenated by adding a little bicarbonate of soda to the cooking water. Adding the bicarb has a double effect: it helps the colour and helps keep them crisp too.

White Cauliflower

Two tablespoons bicarbonate of soda added to the cooking water when you are boiling a cauliflower will help keep the flower of the vegetable nice and white.

Crisp Corn

If you want to keep your corn on the cob crisp, why not try adding 2 tablespoons bicarbonate of soda to the cooking water?

Soaking Beans

If you like to eat dried beans but want to soften them before cooking, you can help them along when soaking them overnight by adding 1 teaspoon bicarbonate of soda to the water – with enough water to cover the beans fully.

Great Greens

To maintain the colour of green vegetables, you can add ½ teaspoon bicarbonate of soda to the cooking water. This works really well for green beans, broccoli and sprouts.

✋ **CAUTION:** Do not forget that you should never overcook your greens – if you do, most of the goodness ends up in the water.

First Class Cabbage

To reduce the cooking time of cabbage and to keep it tender, you just have to add 3 tablespoons bicarbonate of soda to the cooking water. We all agree that there is nothing worse than overcooked cabbage!

Adding bicarbonate of soda to your cabbage's cooking water can also help your family and friends avoid the ill effects of the vegetable's gasses after eating the meal. It must be much better for all concerned, I'm sure you agree!

 CAUTION: Although bicarbonate of soda has a positive visual effect on vegetables, do not be tempted to add too much or get into a habit of using it all the time, as it can reportedly destroy the natural vitamins C and B1 in the veggies.

Speedy Tomato Sauce

If you are making your own tomato sauce or salsa, you can reduce the cooking time by adding some bicarbonate of soda. Add 2 tablespoons bicarbonate of soda for every 900 ml/1½ pts/scant 1 qt cooking water. This can reduce the cooking time by at least 10 minutes.

Sweeter Tomatoes

Adding bicarbonate of soda to the water you are cooking your tomatoes in when preparing your salsa or tomato sauce can also help to neutralize the acid in the tomatoes.

Milder Veg

In order to remove the harsher and stronger flavour from some wild vegetables, add around 1 teaspoon bicarbonate of soda per 1 l/1¾ pts/1 qt water when you boil your vegetables. The colour of the vegetables also becomes brighter.

Clean Veg

You cannot be too careful when it comes to food handling and preparation. Wash your fruit and vegetables in cold water with 2–3 tablespoons bicarbonate of soda added to it. This will help remove some of the impurities that our tap water leaves behind. For more thorough cleaning, the dirt in or on your fruit and vegetables will wash away so much more quickly and easily if you add 100 g/3½ oz/½ cup bicarbonate of soda to the water.

To go further, if you want to be sure your shop-bought vegetables are chemical-free when you cook them, try washing them with 200 g/7 oz/1 cup bicarbonate of soda added to a sinkful of fresh water. More and more of the ready-prepared vegetables and fruits look too good to be true – shiny fruit is often covered with edible wax, but bicarb will get rid of this for you.

Alternatively, you can dab a wet sponge or your vegetable brush into neat bicarbonate of soda and use this to scrub your vegetables. The bicarbonate of soda will remove the dirt, any wax and pesticides. Give everything a thorough rinsing before serving or cooking your vegetables.

Other Uses

Baking Powder

Ever noticed how some recipes call for bicarbonate of soda while others call for baking powder? As discussed in the Introduction, these are two different things. However, you can make your own baking powder by stirring and sifting together two parts cream of tartar to one part bicarb and one part cornflour.

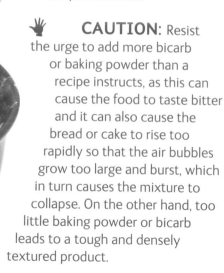

CAUTION: Resist the urge to add more bicarb or baking powder than a recipe instructs, as this can cause the food to taste bitter and it can also cause the bread or cake to rise too rapidly so that the air bubbles grow too large and burst, which in turn causes the mixture to collapse. On the other hand, too little baking powder or bicarb leads to a tough and densely textured product.

A Question of Taste

If you live in a hard-water area, perhaps you don't like the taste of your tap water? To improve the taste you could add 1 teaspoon bicarbonate of soda to each 2.25 l/4 pts/2¼ qts tap water.

Kitchen Cleaning

Don't forget to check out the section on kitchens (pages 14–45) for lots of great tips on keeping your kitchen surfaces, appliances and utensils in tiptop shape with bicarbonate of soda.

Neutralize Acidity

If you suffer with stomach ulcers, it is important that you do not consume too much acid. There is plenty of citric acid in most fruits that can upset and inflame the stomach. You can neutralize the acid in coffee by adding 1 teaspoon bicarbonate of soda to your cup or mug. Why not add it to the coffee pot when you make a fresh brew? Two or 3 teaspoons bicarb will be more than enough for everyone.

No More Beans Blues

You know how the saying goes: 'Beans, beans are good for your heart, the more you eat …' Well, to solve the excess gas problem, try adding 1 teaspoon bicarbonate of soda to your baked beans when you are cooking them. All of the bean goodness is there without the unpleasant aftereffects!

Recipes

Bicarbonate of soda has been used throughout the ages as a raising agent in baking. This is its principal use in food, either used directly or as part of baking powder. If there is already some sort of acid in the recipe, then bicarbonate of soda can be used, as it will react with the acid once a liquid is added and then cause the carbon dioxide bubbles that enable the raising. Baking powder is used when there is no acid in the recipe.

Irish Soda Bread

Makes 1 loaf

400 g/14 oz/3¼ cups plain
white/all-purpose flour,
plus 1 tbsp for dusting

1 tsp salt

2 tsp bicarbonate of
soda/baking soda

15 g/½ oz/1 tbsp butter

50 g/2 oz/⅓ cup coarse
oatmeal/quick oats

1 tsp clear honey

300 ml/½ pt/1¼ cups buttermilk

2 tbsp milk

For a wholemeal variation:

400 g/14 oz/3¼ cups
wholemeal flour,
plus 1 tbsp for dusting

1 tbsp milk

1 Preheat the oven to 200°C/400°F/Gas Mark 6, 15 minutes before baking. Sift the flour, salt and bicarbonate of soda/baking soda into a large bowl. Rub in the butter until the mixture resembles fine breadcrumbs. Stir in the oatmeal and make a well in the centre.

2 Mix the honey, buttermilk and milk together and add to the dry ingredients. Mix to a soft dough.

3 Knead the dough on a lightly floured surface for 2–3 minutes until the dough is smooth. Shape into a 20.5 cm/8 inch round and place on an oiled baking sheet.

4 Thickly dust the top of the bread with flour. Using a sharp knife, cut a deep cross on top, going about halfway through the loaf.

5 Bake in the preheated oven on the middle shelf for 30–35 minutes until the bread is slightly risen, golden and sounds hollow when tapped underneath. Cool on a wire rack. Eat on the day of making.

6 For a wholemeal soda bread, use all the wholemeal flour instead of the white flour and add an extra tablespoon of milk when mixing together. Dust the top with wholemeal flour and bake.

Traditional Oven Scones

Makes 8

225 g/8 oz/1¾ cups
self-raising flour
1 tsp baking powder
pinch salt
40 g/1½ oz/3 tbsp butter, cubed

15 g/½ oz/1 tbsp caster/
superfine granulated sugar
150 ml/¼ pt/⅔ cup milk, plus
1 tbsp for brushing
1 tbsp plain/all-purpose flour,
for dusting

For a lemon and sultana variation:

50 g/2 oz/⅓ cup sultanas/
golden raisins
finely grated zest of ½ lemon
beaten egg, to glaze

1 Preheat the oven to 220°C/425°F/Gas Mark 7, 15 minutes before baking. Sift the flour, baking powder and salt into a large bowl. Rub in the butter until the mixture resembles fine breadcrumbs. Stir in the sugar and mix in enough milk to give a fairly soft dough.

2 Knead the dough on a lightly floured surface for a few seconds until smooth. Roll out until 2 cm/¾ in thick and stamp out 6.5 cm/2½ inch rounds with a floured plain cutter.

3 Place on an oiled baking sheet and brush the tops with milk (do not brush it over the sides or the scones will not rise properly). Dust with a little plain flour.

4 Bake in the preheated oven for 12–15 minutes until well risen and golden brown. Transfer to a wire rack and serve warm or leave to cool completely. (The scones are best eaten on the day of baking, but may be kept in an airtight tin for up to 2 days.)

5 For lemon and sultana scones, stir in the sultanas/raisins and lemon zest with the sugar. Roll out until 2 cm/¾ inch thick and cut into 8 fingers 10 x 2.5 cm/4 x 1 inch in size. Bake the scones as before.

Cheese-crusted Potato Scones

Serves 4

200 g/7 oz/1¾ cups
    self-raising flour
3 tbsp wholemeal/
    whole-wheat flour
½ tsp salt

1½ tsp baking powder
25 g/1 oz/¼ stick butter, cubed
5 tbsp milk
175 g/6 oz/¼ cup cold
    mashed potato
freshly ground black pepper

2 tbsp milk, for brushing
40 g/1½ oz/6 tbsp mature
    Cheddar cheese, finely grated
paprika, to dust
basil sprig, to garnish

1 Preheat the oven to 220°C/425°F/Gas Mark 7, 15 minutes before baking. Sift the flours, salt and baking powder into a large bowl. Rub in the butter until the mixture resembles fine breadcrumbs.

2 Stir 4 tablespoons of the milk into the mashed potato and season with black pepper.

3 Add the dry ingredients to the potato mixture, mixing together with a fork and adding the remaining 1 tablespoon of milk if needed.

4 Knead the dough on a lightly floured surface for a few seconds until smooth. Roll out to a 15 cm/6 inch round and transfer to an oiled baking sheet.

5 Mark the scone round into 6 wedges, cutting about halfway through with a small sharp knife.

6 Brush with milk, then sprinkle with the cheese and a faint dusting of paprika.

7 Bake on the middle shelf of the preheated oven for 15 minutes, or until well risen and golden brown.

8 Transfer to a wire rack and leave to cool for 5 minutes before breaking into wedges.

9 Serve warm or leave to cool completely. Once cool, store the scones in an airtight container. Garnish with a sprig of basil and serve split and buttered.

Drop Scones

Makes 18

white vegetable fat,
    for greasing
175 g/6 oz/1⅓ cups
    self-raising flour
1 tsp baking powder

40 g/1½ oz/scant ¼ cup
    caster/superfine sugar
1 medium/large egg
200 ml/7 fl oz/¾ cup milk
butter and syrup, to serve

1 Grease a heavy-based nonstick frying pan or a flat griddle pan with white vegetable fat and heat gently.
2 Sift the flour and baking powder into a bowl, stir in the sugar and make a well in the centre. Add the egg and half the milk and beat to a smooth thick batter. Beat in enough of the remaining milk to give the consistency of thick cream.
3 Drop the mixture on to the hot pan, 1 heaped tablespoon at a time, spacing them well apart. When small bubbles rise to the surface of each scone, flip them over with a spatula and cook for about 1 minute until golden brown.
4 Place on a serving dish and keep warm, covered with a clean cloth, while you cook the remaining mixture. Serve warm with butter and syrup and eat on the day of making.

Fruity Apple Teabread

Cuts into 12 slices

125 g/4 oz/½ cup butter

125 g/4 oz/½ cup soft light
brown/golden brown sugar

275 g/10 oz/2 cups
sultanas/golden raisins

150 ml/¼ pint/⅔ cup apple juice

1 apple, peeled, cored
and chopped

2 eggs, beaten

275 g/10 oz/2¼ cups plain/
all-purpose flour

½ tsp ground cinnamon

½ tsp ground ginger

2 tsp bicarbonate of
soda/baking soda

butter curls, to serve

To decorate:

1 apple, cored and sliced

1 tsp lemon juice

1 tbsp golden/corn
syrup, warmed

1 Preheat the oven to 180°C/350°F/Gas Mark 4. Oil and line the base of a 900 g/2 lb loaf tin/pan with nonstick baking paper.

2 Put the butter, sugar, sultanas/raisins and apple juice in a small saucepan. Heat gently, stirring occasionally, until the butter has melted. Tip into a bowl and leave to cool.

3 Stir in the chopped apple and beaten eggs. Sift in the flour, spices and bicarbonate of soda/baking soda and stir together. Spoon into the prepared loaf tin and smooth the top level with the back of a spoon.

4 Toss the apple slices in lemon juice and arrange on top. Bake in the preheated oven for 50 minutes. Cover with foil to prevent the top from browning too much.

5 Bake for 30–35 minutes until a skewer inserted into the centre comes out clean. Leave in the tin for 10 minutes before turning out to cool on to a wire rack.

6 Brush the top with syrup and leave to cool. Remove the lining paper, cut into thick slices and serve with curls of butter.

Maple, Pecan & Lemon Loaf

Cuts into 12 slices

350 g/12 oz/2¾ cups plain/
　all-purpose flour

1 tsp baking powder

175 g/6 oz/¾ cup butter, cubed

75 g/3 oz/⅓ cup caster/
　superfine sugar

125 g/4 oz/1 cup pecans,
　roughly chopped

3 eggs, 1 tbsp milk

finely grated zest of 1 lemon

5 tbsp maple syrup

For the icing:

75 g/3 oz icing/powdered sugar

1 tbsp lemon juice

25 g/1 oz/¼ cup pecans,
　roughly chopped

1 Preheat the oven to 170°C/325°F/Gas Mark 3, 10 minutes before baking. Lightly oil and line the base of a 900 g/2 lb loaf tin/pan with nonstick baking paper.

2 Sift the flour and baking powder into a large bowl. Rub in the butter until the mixture resembles fine breadcrumbs. Stir in the caster/superfine sugar and pecans.

3 Beat the eggs together with the milk and lemon zest. Stir in the maple syrup. Add to the dry ingredients and gently stir in until mixed thoroughly to make a soft dropping consistency.

4 Spoon the mixture into the prepared tin and level the top with the back of a spoon. Bake on the middle shelf of the preheated oven for 50–60 minutes until the cake is well risen and lightly browned. If a skewer inserted into the centre comes out clean, then the cake is ready.

5 Leave the cake in the tin for about 10 minutes, then turn out and leave to cool on a wire rack. Carefully remove the lining paper.

6 Sift the icing/powdered sugar into a small bowl and stir in the lemon juice to make a smooth icing. Drizzle the icing over the top of the loaf, then scatter with the chopped pecans. Leave to set, slice thickly and serve.

Carrot Cake

Cuts into 8 slices

200 g/7 oz/1⅔ cups plain/
all-purpose flour
½ tsp ground cinnamon
½ tsp freshly grated nutmeg
1 tsp baking powder
1 tsp bicarbonate of soda
150 g/5 oz/⅔ cup dark
muscovado/brown sugar

200 ml/7 fl oz/¾ cup
vegetable oil
3 eggs
225 g/8 oz/1¼ cups carrots,
peeled and roughly grated
50 g/2 oz/½ cup
chopped walnuts

For the icing:

175 g/6 oz/¾ cup cream cheese
finely grated zest of 1 orange
1 tbsp orange juice
1 tsp vanilla extract
125 g/4 oz/1 cup icing/
powdered sugar

1 Preheat the oven to 150°C/300°F/Gas Mark 2, 10 minutes before baking. Lightly oil and line the base of a 15 cm/6 inch deep square cake tin/pan with nonstick baking paper.

2 Sift the flour, spices, baking powder and bicarbonate of soda together into a large bowl. Stir in the dark muscovado/brown sugar and mix together.

3 Lightly whisk the oil and eggs together, then gradually stir into the flour and sugar mixture. Stir well. Add the carrots and walnuts. Mix thoroughly, then pour into the prepared cake tin. Bake in the preheated oven for 1¼ hours, or until light and springy to the touch and a skewer inserted into the centre of the cake comes out clean.

4 Remove from the oven and allow to cool in the tin for 5 minutes before turning out on to a wire rack. Reserve until cold.

5 To make the icing, beat together the cream cheese, orange zest and juice and vanilla extract. Sift in the icing/powdered sugar and stir. When the cake is cold, discard the lining paper, spread the cream cheese icing over the top and serve cut into squares.

Easy Chocolate Cake

Serves 8–10

75 g/3 oz dark chocolate,
  broken into squares
200 ml/7 fl oz/¾ cup milk
250 g/9 oz/1¼ cups dark
  muscovado/dark brown sugar
75 g/3 oz/⅓ cup (6 tbsp) butter,
  softened

2 medium/large eggs, beaten
150 g/5 oz/1 heaped cup
  plain/all-purpose flour
½ tsp vanilla extract
1 tsp bicarbonate of
  soda/baking soda
25 g/1 oz/¼ cup unsweetened
  cocoa powder

For the topping and filling:

125 g/4 oz/½ cup (1 stick)
  unsalted butter
225 g/8 oz/2¼ cups
  icing/powdered sugar, sifted
8 large fresh strawberries, halved
tiny mint sprigs, to decorate

1 Preheat the oven to 180°C/350°F/Gas Mark 4. Grease two 20.5 cm/8 inch round sandwich tins/pans and line the bases with nonstick baking paper.

2 Place the chocolate, milk and 75 g/3 oz/⅓ cup of the sugar in a heavy-based saucepan. Heat gently until the mixture has melted, then set aside to cool.

3 Place the butter and remaining sugar in a large bowl and whisk with an electric mixer until light and fluffy. Gradually whisk in the eggs, adding 1 teaspoon flour with each addition.

4 Stir in the cooled melted chocolate mixture along with the vanilla extract. Sift in the flour, bicarbonate of soda/baking soda and cocoa powder, then fold into the mixture until smooth.

5 Spoon the batter into the tins and smooth level. Bake for about 30 minutes until a skewer inserted into the centre comes out clean. Turn out to cool on a wire rack.

6 To decorate, beat the butter with the icing/powdered sugar and 1 tablespoon warm water until light and fluffy, then place half in a piping bag fitted with a star nozzle.

7 Spread half the buttercream over one sponge layer and scatter half the strawberries over it. Top with the other cake and spread the remaining buttercream over the top. Pipe a border of stars around the edge. Decorate with the remaining strawberries and mint sprigs.

Moist Mocha & Coconut Cake

Makes 9 squares

3 tbsp ground coffee

5 tbsp hot milk

75 g/3 oz/⅔ stick butter

175 g/6 oz/½ cup golden/
   corn syrup

2 tbsp soft light brown sugar

40 g/1½ oz/½ cup desiccated/
   shredded coconut

150 g/5 oz/scant 1¼ cups
   plain/all-purpose flour

25 g/1 oz/⅓ cup cocoa powder
   (unsweetened)

½ tsp bicarbonate of soda/
   baking soda

2 medium/large eggs, beaten

2 chocolate flakes, to decorate

For the coffee icing:

225 g/8 oz/2¼ cups icing/
   powdered sugar, sifted

125 g/4 oz/1 stick plus 1 tbsp
   butter, softened

1 Preheat the oven to 170°C/325°F/Gas Mark 3, 10 minutes before baking. Lightly oil and line a deep 20.5 cm/8 inch square tin/pan with nonstick baking paper. Place the coffee in a small bowl and pour over the hot milk. Leave to infuse for 5 minutes, then strain through a tea-strainer or a muslin-lined sieve. Reserve the resulting 4 tablespoons of liquid.

2 Put the butter, golden/corn syrup, sugar and coconut in a small heavy-based saucepan and heat gently until the butter has melted and the sugar dissolved. Sift the flour, cocoa powder and bicarbonate of soda/baking soda together and stir into the melted mixture with the eggs and 3 tablespoons of the coffee-infused milk. Pour the mixture into the prepared tin. Bake on the centre shelf of the preheated oven for 45 minutes, or until the cake is well risen and firm to the touch. Leave in the tin for 10 minutes to cool slightly, then turn out on to a wire rack to cool completely.

3 For the icing, gradually add the icing/powdered sugar to the softened butter and beat together until mixed. Add the remaining 1 tablespoon of coffee-infused milk and beat until light and fluffy. Carefully spread the coffee icing over the top of the cake, then cut into 9 squares. Decorate each square with a small piece of chocolate flake and serve.

Lemon-iced Ginger Squares

Makes 12

225 g/8 oz/1 cup
   caster/superfine sugar
50 g/2 oz/½ stick butter, melted
2 tbsp black treacle/molasses
2 egg whites, lightly whisked

225 g/8 oz/2 cups plain/all-
   purpose flour
1 tsp bicarbonate of soda/
   baking soda
½ tsp ground cloves
1 tsp ground cinnamon

¼ tsp ground ginger
pinch salt
250 ml/8 fl oz/1 cup buttermilk
175 g/6 oz/1½ cups
   icing/powdered sugar
lemon juice

1 Preheat the oven to 200°C/ 400°F/Gas Mark 6, 15 minutes before baking. Lightly oil a 20.5 cm/8 inch square cake tin/pan and sprinkle with a little flour.

2 Mix together the caster/superfine sugar, butter and treacle/molasses. Stir in the egg whites.

3 Mix together the flour, bicarbonate of soda/baking soda, cloves, cinnamon, ginger and salt.

4 Stir the flour mixture and buttermilk alternately into the butter mixture until well blended.

5 Spoon into the prepared tin and bake in the preheated oven for 35 minutes, or until a skewer inserted into the centre of the cake comes out clean.

6 Remove from the oven and allow to cool for 5 minutes in the tin before turning out on to a wire rack over a large plate. Using a cocktail stick/toothpick, make holes in the top of the cake.

7 Meanwhile, mix together the icing/powdered sugar with enough lemon juice to make a smooth pourable icing.

8 Carefully pour the icing over the hot cake, then leave until cold. Cut the ginger cake into squares and serve.

Honey Cake

Cuts into 6 slices

50 g/2 oz/¼ cup butter

2 tbsp caster/superfine sugar

125 g/4 oz/⅓ cup clear honey

175 g/6 oz/1⅓ cups plain/
all-purpose flour

½ tsp bicarbonate of
soda/baking soda

½ tsp mixed/pumpkin
pie spice

1 medium/large egg

2 tbsp milk

25 g/1 oz/¼ cup flaked almonds

1 tbsp clear honey, to drizzle

1 Preheat the oven to 180°C/350°F/Gas Mark 4, 10 minutes before baking. Lightly oil and line the base of an 18 cm/7 inch deep round cake tin/pan with lightly oiled greaseproof/waxed paper or baking paper.

2 In a saucepan, gently heat the butter, sugar and honey until the butter has just melted.

3 Sift the flour, bicarbonate of soda/baking soda and mixed spice together into a bowl.

4 Beat the egg and the milk until thoroughly mixed.

5 Make a well in the centre of the sifted flour and pour in the melted butter and honey.

6 Using a wooden spoon, beat well, gradually drawing in the flour from the sides of the bowl.

7 When all the flour has been beaten in, add the egg mixture and mix thoroughly. Pour into the prepared tin and sprinkle with the flaked almonds.

8 Bake in the preheated oven for 30–35 minutes until well risen and golden brown and a skewer inserted into the centre of the cake comes out clean.

9 Remove from the oven, cool for a few minutes in the tin before turning out and leaving to cool on a wire rack. Drizzle with the remaining tablespoon of honey and serve.

Gingerbread

Cuts into 8 slices

175 g/6 oz/1½ sticks butter
or margarine
225 g/8 oz/⅔ cup black
treacle/molasses
50 g/2 oz/¼ cup dark
muscovado/dark brown sugar

350 g/12 oz/3 cups plain/
all-purpose flour
2 tsp ground ginger
150 ml/¼ pint/⅔ cup milk,
warmed
2 eggs

1 tsp bicarbonate of
soda/baking soda
1 piece stem ginger in syrup
1 tbsp stem ginger syrup

1 Preheat the oven to 150°C/300°C/Gas Mark 2, 10 minutes before baking. Lightly oil and line the base of a 20.5 cm/8 inch deep round cake tin/pan with greaseproof/waxed paper or baking paper.

2 In a saucepan, gently heat the butter or margarine, black treacle/molasses and sugar, stirring occasionally until the butter melts. Leave to cool slightly.

3 Sift the flour and ground ginger into a large bowl. Make a well in the centre, then pour in the treacle mixture. Reserve 1 tablespoon of the milk, then pour the rest into the treacle mixture. Stir together lightly until mixed.

4 Beat the eggs together, then stir into the mixture.

5 Dissolve the bicarbonate of soda/baking soda in the remaining 1 tablespoon of warmed milk and add to the mixture. Beat the mixture until well mixed and free of lumps.

6 Pour into the prepared tin and bake in the preheated oven for 1 hour, or until well risen and a skewer inserted into the centre comes out clean.

7 Cool in the tin, then remove. Slice the stem ginger into thin slivers and sprinkle over the cake. Drizzle with the syrup and serve.

Easy Victoria Sponge

Serves 8

225 g/8 oz/1 cup soft margarine

225 g/8 oz/1¼ cup
    caster/superfine sugar

4 medium/large eggs

1 tsp vanilla extract

225 g/8 oz/1¾ cups
    self-raising flour

1 tsp baking powder

icing/powdered sugar, to dust

For the filling:

4 tbsp seedless raspberry jam

100 ml/3½ fl oz/⅓ cup
    double/heavy cream

1 Preheat the oven to 180°C/350°F/Gas Mark 4. Grease two 20.5 cm/8 inch sandwich tins/pans and line the bases with nonstick baking paper.

2 Place the margarine, sugar, eggs and vanilla extract in a large bowl and sift in the flour and baking powder. Beat for about 2 minutes until smooth and blended, then divide between the tins and smooth level.

3 Bake for about 25 minutes until golden, well risen and the tops of the cakes spring back when lightly touched with a fingertip. Leave to cool in the tins for 2 minutes, then turn out on to a wire rack to cool. When cold, peel away the baking paper.

4 When completely cold, spread one cake with jam and place on a serving plate. Whip the cream until it forms soft peaks, then spread on the underside of the other cake. Sandwich the two cakes together and sift a little icing/powdered sugar over the top.

Orange Fruit Cake

Cuts into 10–12 slices

225 g/8 oz/1¾ cups
self-raising flour
2 tsp baking powder
225 g/8 oz/1 heaped cup
caster/superfine sugar
225 g/8 oz/2 sticks butter, soft
4 large/extra-large eggs
grated zest of 1 orange

2 tbsp orange juice
2–3 tbsp Cointreau
125 g/4 oz/1 cup chopped nuts
Cape gooseberries, blueberries,
raspberries, mint, to decorate
icing/powdered sugar, to dust

For the filling:

450 ml/¾ pint/1¾ cups
double/heavy cream

50 ml/2 fl oz/¼ cup plain
Greek yogurt
½ tsp vanilla extract
2–3 tbsp Cointreau
1 tbsp icing/powdered sugar
450 g/1 lb orange fruits, such
as mango, peach, nectarine,
papaya and yellow plums

1 Preheat the oven to 180°C/350°F/Gas Mark 4, 10 minutes before baking. Lightly oil and line the base of a 25.5 cm/10 inch deep cake tin/pan with nonstick baking paper.

2 Sift the flour and baking powder into a large bowl and stir in the sugar. Make a well in the centre and add the butter, eggs, grated zest and orange juice. Beat until blended and a smooth batter is formed. Turn into the tin and smooth the top. Bake in the preheated oven for 35–45 minutes until golden and the sides begin to shrink from the edge of the tin. Remove, cool before removing from the tin and discard the lining paper. Using a serrated knife, slice off the top third of the cake, cutting horizontally. Sprinkle the cut sides with the Cointreau.

3 For the filling, whip the cream and yogurt with the vanilla extract, Cointreau and icing/powdered sugar until soft peaks form. Chop the orange fruit and fold into the cream. Spread some of this mixture on to the bottom cake layer. Transfer to a serving plate. Cover with the top layer of sponge and spread the remaining cream mixture over the top and sides. Press the chopped nuts into the sides of the cake and decorate the top with the Cape gooseberries, blueberries and raspberries. Dust the top with icing sugar and serve.

Chunky Chocolate Muffins

Makes 7

50 g/2 oz dark chocolate, roughly chopped

50 g/2 oz/¼ cup light muscovado/golden brown sugar

25 g/1 oz/¼ stick butter, melted

125 ml/4 fl oz/½ cup milk, at room temperature

½ tsp vanilla extract

1 medium/large egg, lightly beaten

150 g/5 oz/1¼ cups self-raising flour

½ tsp baking powder

pinch salt

75 g/3 oz/3 squares white chocolate, chopped

2 tsp icing/powdered sugar (optional)

1 Preheat the oven to 200°C/ 400°F/Gas Mark 6, 15 minutes before baking. Line a muffin tin/pan with 7 paper muffin cases/baking cups or oil the individual compartments well. Place the dark chocolate in a large heatproof bowl set over a saucepan of very hot water and stir occasionally until melted. Remove the bowl and leave to cool for a few minutes.

2 Stir the sugar and butter into the melted chocolate, then the milk, vanilla extract and egg. Sift in the flour, baking powder and salt together. Add the chopped white chocolate, then, using a metal spoon, fold together quickly, taking care not to overmix.

3 Divide the mixture between the paper cases, piling it up in the centre. Bake on the centre shelf of the preheated oven for 20–25 minutes until well risen and firm to the touch.

4 Lightly dust the tops of the muffins with icing/powdered sugar, if using, as soon as they come out of the oven. Leave the muffins in the tins for a few minutes, then transfer to a wire rack. Serve warm or cold.

Chocolate Chip Cherry Muffins

Makes 12

75 g/3 oz/⅓ cup glacé/
candied cherries

75 g/3 oz/scant ½ cup
milk/semisweet or dark
chocolate chips

75 g/3 oz/⅓ cup soft margarine

200 g/7 oz/1 cup
caster/superfine sugar

2 medium/large eggs

150 ml/¼ pint/⅔ cup thickset
natural yogurt

5 tbsp milk

275 g/10 oz/2 heaped cups
plain/all-purpose flour

1 tsp bicarbonate of
soda/baking soda

1 Preheat the oven to 200°C/400°F/Gas Mark 6. Line a deep 12-hole muffin tray with deep paper cases/baking cups. Wash and dry the cherries. Chop them roughly, mix them with the chocolate chips and set aside.

2 Beat the margarine and sugar together, then whisk in the eggs, yogurt and milk. Sift in the flour and bicarbonate of soda/baking soda. Stir until just combined.

3 Fold in three quarters of the cherries and chocolate chips. Spoon the mixture into the cases, filling them two-thirds full. Sprinkle the remaining cherries and chocolate chips over the top.

4 Bake for about 20 minutes until golden and firm. Leave in the tins for 4 minutes, then turn out to cool on a wire rack.

Chocolate Chip Cookies

Makes 36 cookies

175 g/6 oz/1½ cups plain/
   all-purpose flour
pinch salt
1 tsp baking powder
¼ tsp bicarbonate of
   soda/baking soda

75 g/3 oz/6 tbsp butter
   or margarine
50 g/2 oz/4 tbsp soft light
   brown sugar
3 tbsp golden/corn syrup
125 g/4 oz/¾ cup
   chocolate chips

1 Preheat the oven to 190°C/375°F/Gas Mark 5, 10 minutes before baking. Lightly oil a large baking sheet.

2 In a large bowl, sift together the flour, salt, baking powder and bicarbonate of soda/baking soda.

3 Cut the butter or margarine into small pieces and add to the flour mixture.Using 2 knives or your fingertips, rub in the butter or margarine until the mixture resembles coarse breadcrumbs. Add the light brown sugar, golden/corn syrup and chocolate chips. Mix together until a smooth dough forms.

4 Shape the mixture into small balls and arrange on the baking sheet, leaving enough space to allow them to expand. (These cookies do not increase in size by a great deal, but allow a little space for expansion.) Flatten the mixture slightly with the fingertips or the heel of the hand. Bake in the preheated oven for 12–15 minutes until golden and cooked through.

5 Allow to cool slightly, then transfer the biscuits on to a wire rack to cool. Serve when cold or otherwise store in an airtight container.

White Chocolate Cookies

Makes about 24

140 g/5 oz/⅔ cup butter

40 g/1½ oz/3 tbsp caster/superfine sugar

60 g/2½ oz/⅓ cup soft dark brown sugar

1 egg

125 g/4 oz/1 cup plain/ all-purpose flour

½ tsp bicarbonate of soda/baking soda

few drops vanilla extract

150 g/5 oz white chocolate

50 g/2 oz/½ cup whole hazelnuts, shelled

1 Preheat the oven to 180°C/350°F/Gas Mark 4, 10 minutes before baking. Lightly butter several baking sheets with a tablespoon of the butter. Place the remaining butter with both sugars into a large bowl and beat with a wooden spoon or an electric mixer until soft and fluffy.

2 Beat the egg, then gradually mix into the creamed mixture. Sift the flour and the bicarbonate of soda/baking soda together, then carefully fold into the creamed mixture with a few drops of vanilla extract.

3 Roughly chop the chocolate and hazelnuts into small pieces, add to the bowl and gently stir into the mixture. Mix together lightly to blend.

4 Spoon heaped teaspoons of the mixture on to the prepared baking sheets, making sure that there is plenty of space in between each one as they will spread a lot during cooking.

5 Bake the cookies in the preheated oven for 10 minutes, or until golden, then remove from the oven and leave to cool for 1 minute. Using a spatula, carefully transfer to a wire rack and leave to cool completely. The cookies are best eaten on the day they are made. Store in an airtight container.

Chocolate-covered Flapjacks

Makes 24

215 g/7½ oz/1¾ cups plain/
all-purpose flour
150 g/5 oz/1¼ cups rolled oats
225 g/8 oz/1 cup light muscovado
/golden brown sugar

1 tsp bicarbonate of soda/
baking soda
pinch salt
150 g/5 oz/⅔ cup butter
2 tbsp golden/corn syrup

250 g/9 oz dark chocolate
5 tbsp double/heavy cream

1 Preheat the oven to 180°C/350°F/Gas Mark 4, 10 minutes before baking. Lightly oil a 33 x 23 cm/13 x 9 inch Swiss-/jelly-roll tin and line with nonstick baking paper. Place the flour, rolled oats, light muscovado/golden brown sugar, bicarbonate of soda/baking soda and salt into a bowl and stir well together.

2 Melt the butter and golden/corn syrup together in a heavy-based saucepan and stir until smooth, then add to the oat mixture and mix together thoroughly. Spoon the mixture into the prepared tin, press down firmly and level the top.

3 Bake in the preheated oven for 15–20 minutes until golden. Remove from the oven and leave the flapjack to cool in the tin. Once cool, remove from the tin. Discard the paper.

4 Melt the chocolate in a heatproof bowl set over a saucepan of gently simmering water. Alternatively, melt the chocolate in the microwave according to the manufacturer's instructions. Once the chocolate has melted, quickly beat in the cream, then pour over the flapjack. Mark patterns over the chocolate with a fork when almost set.

5 Chill the flapjack in the refrigerator for at least 30 minutes before cutting into bars. When the chocolate has set, serve. Store in an airtight container for a few days.

Ginger Snaps

Makes 40

300 g/11 oz/1⅓ cups butter or
  margarine, softened
225 g/8 oz/1 cup soft light
  brown sugar
75 g/3 oz/3 tbsp black
  treacle/molasses

1 egg
400 g/14 oz/3¼ cups plain/
  all-purpose flour
2 tsp bicarbonate of
  soda/baking soda
½ tsp salt
1 tsp ground ginger

1 tsp ground cloves
1 tsp ground cinnamon
50 g/2 oz/4 tbsp
  granulated sugar

1 Preheat the oven to 190°C/375°F/Gas Mark 5, 10 minutes before baking. Lightly oil a baking sheet.

2 Cream together the butter or margarine and the sugar until light and fluffy.

3 Warm the treacle/molasses in the microwave for 30–40 seconds, then add gradually to the butter mixture with the egg. Beat until well combined.

4 In a separate bowl, sift the flour, bicarbonate of soda/baking, salt, ground ginger, ground cloves and ground cinnamon. Add to the butter mixture and mix together to form a firm dough.

5 Chill in the refrigerator for 1 hour. Shape the dough into small balls and roll in the granulated sugar. Place well apart on the oiled baking sheet.

6 Sprinkle the baking sheet with a little water and transfer to the preheated oven.

7 Bake for 12 minutes until golden and crisp. Transfer to a wire rack to cool and serve.

Gingerbread Biscuits

Makes 20 large or 28 small biscuits

225 g/8 oz/1¾ cups plain/all-purpose flour, plus extra for dusting

½ tsp ground ginger

½ tsp mixed spice

½ tsp bicarbonate of soda/baking soda

75 g/3 oz/⅓ cup (6 tbsp) butter

2 tbsp golden/corn syrup

1 tbsp black treacle/molasses

75 g/3 oz/⅓ cup soft dark brown sugar

50 g/2 oz/scant ½ cup royal icing sugar, to decorate

1 Preheat the oven to 180°C/350°F/Gas Mark 4 and grease two baking sheets. Sift the flour, spices and bicarbonate of soda/baking soda into a bowl.

2 Place the butter, syrup, treacle/molasses and sugar in a heavy-based pan with 1 tablespoon water and heat gently until every grain of sugar has dissolved and the butter has melted. Cool for 5 minutes, then pour the melted mixture into the dry ingredients and mix to a soft dough.

3 Leave the dough, covered, for 30 minutes. Roll out the dough on a lightly floured surface to a 3 mm/⅛ inch thickness and cut out fancy shapes. Gather up the trimmings and re-roll the dough, cutting out more shapes. Place on the baking sheets using a spatula and bake for about 10 minutes until golden and firm. Be careful not to overcook, as the biscuits will brown quickly.

4 Decorate the biscuits by mixing the royal icing sugar with enough water to make a piping consistency. Place the icing in a small paper piping bag with the end snipped away and pipe faces and decorations on to the biscuits.

Pumpkin Cookies with Brown Butter Glaze

Makes 48

125 g/4 oz/½ cup butter,
  softened
150 g/5 oz/1¼ cups plain/
  all-purpose flour
175 g/6 oz/¾ cup soft light
  brown sugar, lightly packed
225 g/8 oz/1 cup canned or
  cooked pumpkin

1 egg, beaten
2 tsp ground cinnamon
2½ tsp vanilla extract
½ tsp baking powder
½ tsp bicarbonate of soda
½ tsp freshly grated nutmeg
125 g/4 oz/1 scant cup
  wholemeal flour

75 g/3 oz/¾ cup pecans,
  roughly chopped
100 g/3½ oz/½ cup raisins
50 g/2 oz/¼ cup unsalted butter
225 g/8 oz/2 cups
  icing/powdered sugar
2 tbsp milk

1 Preheat the oven to 190°C/375°F/Gas Mark 5, 10 minutes before baking. Lightly oil a baking sheet and reserve.

2 Using an electric mixer, beat the butter until light and fluffy. Add the flour, sugar, pumpkin and beaten egg and beat with the mixer until mixed well. Stir in the ground cinnamon, 1 teaspoon of the vanilla extract and then sift in the baking powder, bicarbonate of soda and grated nutmeg. Beat the mixture until well combined, scraping down the sides of the bowl. Add the wholemeal flour, chopped nuts and raisins to the mixture and fold in with a metal spoon or rubber spatula until mixed thoroughly together.

3 Place teaspoonfuls about 5 cm/2 inches apart on to the baking sheet. Bake in the pre-heated oven for 10–12 minutes until the cookie edges are firm.

4 Remove the biscuits from the oven and leave to cool on a wire rack. Meanwhile, melt the butter in a small saucepan over a medium heat until pale and just turning golden brown.

5 Remove from the heat. Add the sugar, remaining vanilla extract and milk, stirring. Drizzle over the cooled cookies and serve.

Chocolate & Nut Refrigerator Biscuits

Makes 18

165 g/5½ oz/1½ sticks slightly
  salted butter

150 g/5 oz/⅔ cup soft dark
  brown sugar

25 g/1 oz/2 tbsp
  granulated sugar

1 medium egg, beaten

200 g/7 oz/1¾ cups plain flour

½ tsp bicarbonate of
  soda/baking soda

25 g/1 oz/¼ cup cocoa powder

125 g/4 oz/1 cup pecan nuts,
  finely chopped

1 Cream 150 g/5 oz of the butter and both sugars in a large bowl until light and fluffy, then gradually beat in the egg.

2 Sift the flour, bicarbonate of soda/baking soda and cocoa powder together, then gradually fold into the creamed mixture together with the chopped pecans. Mix thoroughly until a smooth but stiff dough is formed.

3 Place the dough on a lightly floured surface or pastry board and roll into sausage shapes about 5 cm/2 inches in diameter. Wrap in clingfilm/plastic wrap and chill in the refrigerator for at least 12 hours, or preferably overnight.

4 Preheat the oven to 190°C/375°F/Gas Mark 5, 10 minutes before baking. Lightly grease several baking sheets with the remaining butter. Cut the dough into thin slices and place on the prepared baking sheets. Bake in the preheated oven for 8–10 minutes until firm. Remove from the oven and leave to cool slightly. Using a spatula, transfer to a wire rack to cool. Store in an airtight container.

Oatmeal Raisin Cookies

Makes 24

175 g/6 oz/1½ cups plain/all-purpose flour

150 g/5 oz/2 cups rolled oats

1 tsp ground ginger

½ tsp baking powder

½ tsp bicarbonate of soda/baking soda

125 g/4 oz/⅔ cup demerara/turbinado sugar

50 g/2 oz/⅓ cup raisins

1 medium/large egg, lightly beaten

150 ml/¼ pint/⅔ cup vegetable or sunflower/corn oil

4 tbsp milk

1 Preheat the oven to 200°C/400°F/Gas Mark 6, 15 minutes before baking. Lightly oil a baking sheet.

2 Mix together the flour, oats, ground ginger, baking powder, bicarbonate of soda/baking soda, sugar and the raisins in a large bowl.

3 In another bowl, mix the egg, oil and milk together. Make a well in the centre of the dry ingredients and pour in the egg mixture.

4 Mix the mixture together well with either a fork or a wooden spoon to make a soft but not sticky dough.

5 Place spoonfuls of the dough well apart on the oiled baking sheet and flatten the tops down slightly with the tines of a fork.

6 Transfer the biscuits to the preheated oven and bake for 10–12 minutes until golden.

7 Remove from the oven, leave to cool for 2–3 minutes, then transfer the biscuits to a wire rack to cool. Serve when cold or otherwise store in an airtight container.

Oatmeal Coconut Cookies

Makes 40

225 g/8 oz/2 sticks butter
or margarine

125 g/4 oz/⅔ cup demerara/
turbinado sugar

125 g/4 oz/⅔ cup caster/
superfine sugar

1 large/extra-large egg,
lightly beaten

1 tsp vanilla extract

225 g/8 oz/2 cups plain/
all-purpose flour

1 tsp baking powder

½ tsp bicarbonate of
soda/baking soda

125 g/4 oz/1¼ cups rolled oats

75 g/3 oz/1 cup desiccated/
shredded coconut

1 Preheat the oven to 180°C/350°F/Gas Mark 4, 10 minutes before baking. Lightly oil a baking sheet.

2 Cream together the butter or margarine and sugars until light and fluffy.

3 Gradually stir in the egg and vanilla extract and beat until well blended.

4 Sift together the flour, baking powder and bicarbonate of soda/baking soda in another bowl.

5 Add to the butter and sugar mixture and beat together until smooth. Fold in the rolled oats and coconut with a metal spoon or rubber spatula.

6 Roll heaped teaspoonfuls of the mixture into balls and place on the baking sheet about 5 cm/2 inches apart and flatten each ball slightly with the heel of the hand.

7 Transfer to the preheated oven and bake for 12–15 minutes, until just golden.

8 Remove from the oven and transfer the biscuits to a wire rack to cool completely and serve.

Honey & Chocolate Hearts

Makes about 20

65 g/2½ oz/scant ⅓ cup
caster/superfine sugar
1 tbsp butter
125 g/4 oz/⅓ cup thick honey
1 small egg, beaten

pinch salt
1 tbsp mixed/candied peel or
chopped glacé/candied
ginger
¼ tsp ground cinnamon
pinch ground cloves

225 g/8 oz/1¾ cups plain/all-
purpose flour, sifted
½ tsp baking powder, sifted
75 g/3 oz milk/semisweet
chocolate

1 Preheat the oven to 220°C/425°F/Gas Mark 7, 15 minutes before baking. Lightly oil two baking sheets. Heat the sugar, butter and honey together in a small saucepan until everything has melted and the mixture is smooth.

2 Remove from the heat and stir until slightly cooled, then add the beaten egg with the salt and beat well. Stir in the mixed/candied peel or glace/candied ginger, ground cinnamon, ground cloves, the flour and the baking powder and mix well until a dough is formed. Wrap in clingfilm/plastic wrap and chill in the refrigerator for 45 minutes.

3 Place the chilled dough on a lightly floured surface, roll out to about 5 mm/¼ inch thickness and cut out small heart shapes. Place on to the prepared baking sheets and bake in the preheated oven for 8–10 minutes. Remove from the oven and leave to cool slightly. Using a spatula, transfer to a wire rack until cold.

4 Melt the chocolate in a heatproof bowl set over a saucepan of simmering water. Alternatively, melt the chocolate in the microwave according to the manufacturer's instructions, until smooth. Dip one half of each biscuit in the melted chocolate. Leave to set before serving.

Chocolate Orange Biscuits

Makes 30

100 g/3½ oz dark chocolate

125 g/4 oz/1 stick plus 1 tbsp
  butter

125 g/4 oz/⅔ cup
  caster/superfine sugar

pinch salt

1 egg, beaten

grated zest of 2 oranges

200 g/7 oz/1½ cups plain/all-
  purpose flour

1 tsp baking powder

100 g/3½ oz/1 cup
  icing/powdered sugar

1–2 tbsp orange juice

1 Preheat the oven to 200°C/400°F/Gas Mark 6, 15 minutes before baking. Lightly oil several baking sheets. Coarsely grate the chocolate and reserve. Beat the butter and sugar together until creamy. Add the salt, beaten egg and half the orange zest and beat again.

2 Sift the flour and baking powder, add to the bowl with the grated chocolate and beat to form a dough. Shape into a ball, wrap in clingfilm/plastic wrap and chill in the refrigerator for 2 hours.

3 Roll the dough out on a lightly floured surface to 5 mm/¼ inch thickness and cut into 5 cm/2 inch rounds. Place the rounds on the prepared baking sheets, allowing room for expansion. Bake in the preheated oven for 10–12 minutes until firm. Remove the biscuits from the oven and leave to cool slightly. Using a spatula, transfer to a wire rack and leave to cool.

4 Sift the icing/powdered sugar into a small bowl and stir in sufficient orange juice to make a smooth, spreadable icing. Spread the icing over the biscuits, leave until almost set, then sprinkle on the remaining grated orange zest before serving.

Rum & Chocolate Squares

Makes 14–16

125 g/4 oz/1 stick plus
  1 tbsp butter
100 g/3½ oz/½ cup caster/
  superfine sugar
pinch salt

2 medium/large egg yolks
225 g/8 oz/1¾ cups plain/all-
  purpose flour
50 g/2 oz/scant ½ cup
  cornflour/cornstarch
¼ tsp baking powder

2 tbsp cocoa powder
  (unsweetened)
1 tbsp rum

1 Preheat the oven to 190°C/375°F/Gas Mark 5, 10 minutes before baking. Lightly oil several baking sheets. Cream the butter, sugar and salt together in a large bowl until light and fluffy. Add the egg yolks and beat well until smooth.

2 Sift together 175 g/6 oz of the flour, the cornflour and the baking powder and add to the mixture and mix well with a wooden spoon until a smooth and soft dough is formed.

3 Halve the dough and knead the cocoa powder into one half and the rum and the remaining plain flour into the other half. Place the two mixtures in two separate bowls, cover with clingfilm/plastic wrap and chill in the refrigerator for 1 hour.

4 Roll out both pieces of dough separately on a well-floured surface into two thin rectangles. Place one on top of the other, cut out squares approximately 5 cm/2 inches x 5 mm/¼ inch and place on the prepared baking sheets.

5 Bake in the preheated oven, half with the chocolate uppermost and the other half rum-side up, for 10–12 minutes until firm. Remove from the oven and leave to cool slightly. Using a spatula, transfer to a wire rack and leave to cool, then serve.

Further Reading

Briggs, M., *Bicarbonate of Soda: A Very Versatile Natural Substance*, Black and White Publishing, 2007

Briggs, M., *Green Cleaning: Natural Hints and Tips*, Abbeydale Press, 2008

Briggs, M., *Vinegar – 1001 Practical Uses*, Abbeydale Press, 2006

Constantino, M., *Household Hints*, Flame Tree, 2009

Fraser, R., *Neal's Yard Remedies Recipes for Natural Beauty*, Haldane Mason Ltd, 2007

Grace, J. L., *Imperfectly Natural Home: The Organic Bible*, Orion, 2008

Hamilton, A., *The Self Sufficient-ish Bible*, Hodder & Stoughton, 2009

Harrison, J., *Low-Cost Living: Live better, spend less*, Right Way, 2009

Lansky, V., *Baking Soda: Over 500 Fabulous, Fun, and Frugal Uses You've Probably Never Thought of*, Book Peddlers, 2008

Logan, K.N., *Clean House, Clean Planet: Clean Your House for Pennies a Day, the Safe, Nontoxic Way*, Simon & Schuster, 1997

Martin, A., *Natural Stain Remover: Clean Your Home Without Harmful Chemicals*, Apple Press, 2003

Peacock, D., *Good Home Cooking: Make it, Don't Buy It! Real Food at Home – Mostly at Less Than a Pound a Head*, Spring Hill 2009

Peacock, P., *Grandma's Ways for Modern Days: Relearning Traditional Self-sufficiency – Gardening, Cooking and Household Management*, Spring Hill, 2009

Reader's Digest, *Extraordinary Uses For Ordinary Things: 2,209 Ways to Save Money and Time*, Reader's Digest, 2008

Reader's Digest, *1001 Home Remedies: Trustworthy Treatments for Everyday Health Problems*, Reader's Digest, 2005

Strawbridge, D., & Strawbridge, J., *Practical Self Sufficiency: The Complete Guide to Sustainable Living*, Dorling Kindersley, 2010

Websites

www.armhammer.com
The official website for the Arm & Hammer baking soda company. It offers practical uses for bicarb, including a section on projects for kids.

www.babycentre.co.uk
A website that offers plenty of advice about caring for babies from pregnancy to preschool.

www.bbc.co.uk/food/campaigns/get-baking
The BBC encourages readers online to 'get baking' by offering various recipes – many of which contain bicarbonate of soda.

www.channel4.com/food/recipes/baking
A website that offers many baking ideas, including many recipes that involve bicarbonate of soda.

www.doityourself.com
A website that offers various DIY tips and projects.

www.enjo.org.uk
A website that offers chemical-free, environmentally friendly cleaning products for all areas of the home.

www.hygieneexpert.co.uk
A website that offers hygiene advice and information for humans as well as pets.

www.gardenersworld.com
A website with tips and guides for gardeners of any experience level.

www.grist.org
An environmentally conscious news site that gives tips on DIY projects and green living.

www.home-remedies-for-you.com
A website that offers treatments for a plethora of ailments using common household products, including good old bicarb.

www.treehugger.com
A website that focuses entirely on green living. It contains news about the latest environmentally friendly technology as well as advice on how to make a positive environmental impact.

www.ultimatehandyman.co.uk
A website that offers advice on DIY home improvement and gives instructions on undertaking many household DIY projects.

www.wackyuses.com
A website that offers unusual but helpful uses for common household products.

Picture Credits

The following images are © **Foundry Arts**: 259, 263 (both Paul Forrester), 288, and 291–345. All other images are courtesy of **Shutterstock** and © the following photographers: 1 & 249, 4t & 46 & 53, 218, 275 Monkey Business Images; 3 & 169, 170 auremar; 48, 144–45 digieye; 4b & 191 Karen H. Ilgan; 4c & 117 Anthony Harris; 55, 81 Baloncici; 5c & 247 HP_photo; 5t & 219 Gemenacom; 6 & 146 & 157, 50, 134, 179 Diego Cervo; 7 Andre Klaassen; 8, 127 Picsfive; 9, 12–13 matka_Wariatka; 10 Jonathan Vasata; 11 & 167 Kurhan; 14 & 23, 175 Julija Sapic; 15 oleksa ; 16 Victoria Alexandrova; 17, 18 Photoroller; 19, 257 Irina Fischer; 20 margouillat photo; 21 Tom Baker; 22 Chris Harvey; 24 artproem; 25 M.E. Mulder; 26 Daniel Krylov; 27 James M Phelps Jr; 29 Ragne Kabanova; 30 UlianaSt; 31 jokerpro; 32 Michael Perrigrew; 33 Anne Kitzman; 35 Tomasz Trojanowski; 36 HABRDA; 37 Olivier Le Queinec; 38 Pavelk; 39 Paul Cowan; 40 sjeacle; 41 Muriel Losure; 42 Lepas, 43 Cheryl Casey; 44 SasPartout; 45 Vadym Drobot; 47 yampi; 48 Brian A Jackson; 51 Crystal Kirk; 52 Tiplyashin Anatoly; 54 Alistair Scott; 56 Freerk Brouwer; 57 Yellowj; 58 Paul Maguire; 59 Daniel Goodings; 60 & 63, 64 Konovalikov Andrey; 61 Gorin; 62 aceshot1; 65 terekhov igor; 66 Margo Harrison; 67 michaeljung; 68 Kimberly Hall; 69, 119 Peter Gudella; 70, 149 jkitan; 71 Joe Belanger; 72 Lori Sparkia; 73, 213 Jaimie Duplass; 75 Terry Underwood Evans; 76 Freddy Eliasson; 77 Shane White; 78 Yuri Arcurs; 79 Olga Chernetskaya; 80 khz; 82 Darren A Hubley; 83 Maria Dryfhout; 85 Leifstiller; 86 Alkristeena; 87 PAUL ATKINSON; 88 Matt Ragen; 89, 193 Zholobov Vadim; 90 & 349 Germany Feng; 91 Supri Suharjoto; 92 Paul Matthew Photography; 93 Lisa Turay; 95 ushama; 96 Romanchuk Dimitry; 97 Maridav; 99 Nagy Jozsef - Attila; 100tl Marie C. Fields; 100br trailexlorers; 101 photogl; 102 & 109 Carrieanne Larmore; 103 Vuk Nenezic; 104 Baevskiy Dmitry; 105 & 347 Michelle D Milliman; 106 iofoto; 107 Slasha; 108 Onur ERSIN; 110tl Joy Brown; 110br Tootles; 111 Sarah Salmela; 112 Thomas Skjæveland; 113 MattJones; 114–15 ZTS; 116 & 123 pixelman; 118 Toranico; 120 Shantell photographe; 121, 228 teekaygee; 122 Joseph; 124 Igor Dutina; 125 Eric Gevaert; 126 Paul Krugloff; 128 Karkas; 129 Petro Feketa; 130 Danylchenko Iaroslav; 131 Agb; 132 Martin Smith; 133 Babusi Octavian Florentin; 135 Jean Valley; 136 Dan Thomas Brostrom; 137 Mirco Vacca; 138 James Coleman; 139 easyshoot; 140 & 142 Stephen VanHorn; 141 Andersr; 143 Dmitriy Yakovlev; 147 rebvt; 148, 289 Piotr Marcinski; 150 Rene Jansa; 151 Arie v.d. Wolde; 152 pzAxe; 153 Skazka Grez; 154 Danijel Micka; 155 Liv friis-larsen; 156 Poznyakov; 158 Tania Zbrodko; 159 Nicolas Dufresne; 160 Subbotina Anna ; 161 Anthony Berenyi; 162 Alexey Stiop; 163 Netfalls; 164 imageshunter; 165 kristian sekulic; 166 Camilo Torres; 168 Alexa Catalin; 171 StockLite; 172 paul prescott; 173 Levent Konuk; 174 sevenke; 176 Natthawat Wongrat ; 177, 194 & 199 Elena Elisseeva; 178 GraÃ§a Victoria ; 180 & 189 Vasina Natalia; 181 elaine hudson; 182 charles taylor; 183 Wojciech Zbieg; 184 Gaby Kooijman; 185 Artur Bogacki; 186 Kristy Pargeter; 187 Zsolt Nyulaszi; 188r James R T Bossert; 188l Kafer photo; 192, 244 & 261 Ieva Geneciviene; 195 Edw; 196 gsplanet; 197 Miodrag Gajic; 198 Melinda Fawver; 200 Jirsak; 201, 267 Noam Armonn; 202 Joe Gough; 203 Tobik; 204–205 efirm; 206 & 209 Vasily Mulyukin; 207 Scott L. Williams; 210 david n madden; 211 kreego; 212 Nancy Kennedy; 214 Cristi Lucaci; 215 Kimmit; 216 Cherick; 217 Stuart Monk; 220 dendong; 221 Alena Ozerova; 222 Alena Brozova; 223 Pakhnyushcha; 224 Ivonne Wierink; 225 Vakhrushev Pavel; 226 Johanna Goodyear; 227 Andrey Pavlov; 228 Vaclav Volrab; 230 Ramona Heim; 231 Mona Makela; 232 Christina Richards; 233 Losevsky Pavel; 234 & 241 Dmitry Naumov; 235 SNEHIT; 236 Kathy Piper; 237 Jason Stitt; 238 Pedro Salaverria; 239 yazan masa; 240 Matej Pavlansky; 242–43 Kruchankova Maya; 245 Aleksandar-Pal Sakala; 246 Alex Staroseltsev; 248, 254, 284 Thomas M Perkins; 250 Yurchyks; 251 Serghei Starus; 253 Robert Anthony; 255 Renata Osinska; 258 Morgar; 260 travis manley; 262 bierchen; 264 Roxana Bashyrova; 265 Whitechild; 266 holbox; 268–69, 278tr Magdalena Kucova; 270 & 283 Tyler Olson; 271 Ermes; 272 Jacek Chabraszewski; 273 David P. Smith; 276 Richard Griffin; 277 Rey Kamensky; 278bl Mark Plumley; 279 jathys; 280 Aleksandra Duda; 281 Denis and Yulia Pogostins; 285 Elke Dennis; 286 Valeriy Velikov; 287 Viktor1.

Index

Recipes have uppercase initials.